THE SAVVY CORPORATE INNOVATOR

KEY STRATEGIES TO GET YOUR BIG IDEA FUNDED IN 30 DAYS

Dorian Simpson

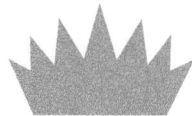

KINGSLEY
PUBLISHING
GROUP

PORTLAND, OREGON

The Savvy Corporate Innovator
Key Strategies to Get Your Big Idea Funded in 30 Days
© 2015 Dorian Simpson

Printed in the United States of America

Editor: Caralee Anley

Kingsley Press
Portland, Oregon
http://www.planninginnovations.com/

ISBN: 978-0-9912201-0-6

Dedicated to Benjamin and Marlowe Simpson.

Thank you for allowing me time away from you to write.
May you lead innovative lives.

CONTENTS

PART I

EXPLORING THE INNOVATOR'S WORLD

PART II

A CORPORATE INNOVATION TRANSLATION

PART III

THE 30-DAY ACTION PLAN

PART IV

LIFE AFTER 30 DAYS

ACKNOWLEDGEMENTS

I cannot express enough thanks to my friends and colleagues who provided support, valuable feedback, and stories for this book. I hope you find the results worthy of your name and energy.

Susannah Axelrod, Tom Carhart, Richard Clem, Eric Colsman, Stephen Davis, Ed de la Fuente, Richard Fast, Greg Gudorf, Roger Hicks, Angela Kienholz, Larry Logan, Doug Means, Steve Morris, Bill Newman, Stan Olson, David Petrie, Michael Reardon, Scott Rosenberg, Edgar Simpson, Wendy Simpson, Clarke Stevens, Marc Tayer, Miki Tokola, Sarah Wischmeyer, David Wood, and Jeff Wolking.

I'd also like to convey a special thanks to Julie Fast who helped me clarify many of the key concepts in this book and made unlimited attempts to teach me how to write a coherent sentence.

Finally, I'd also like to thank all of the unnamed or obscured innovators and managers who have lent their stories to this book for the benefit of my readers.

PREFACE

For my first product idea at Motorola, I spent three months building a detailed opportunity proposal with graphs, multiple scenarios, market research, and financial information for the full funding of the project. At the funding meeting, the general manager flipped through the pages and stopped on the financial statement. "You're showing a gross margin of 30 percent. I don't want to see anything with less than 40 percent." Despite my preparation, I was not ready with a response. This taught me two critical lessons. First, know the exact needs of your audience when presenting an idea. Second, don't waste time developing information or materials that aren't necessary.

Since those early days, I've learned many more lessons and figured out how to move an idea beyond that first spark of excitement. We innovators are inspired thinkers! We come from all disciplines and see opportunities for new products, enhancements, applications or technology, and even new business approaches. We might be in research and development, product management, sales, marketing, operations, or any other function. We play a role in the company, but we mainly want to see our ideas spring into reality.

Although many of the concepts in this book are valid for entrepreneurs, it's primarily for those who work inside established companies and aspire to share ideas. Corporate innovators have unique challenges and must do their work in a company setting with all of its rules and politics. To have success navigating those challenges, innovators can't just have a BIG idea. They need the right, SAVVY skills to develop and present proposals that get the funding to move those ideas forward. My goal is to give you the tools to become a "savvy innovator."

—Dorian

USING THIS BOOK

The Savvy Corporate Innovator focuses on the first 30 days in the life of your BIG idea. It's designed to help you, a savvy innovator, develop a concise, compelling opportunity proposal that gains the support of corporate executives. It will prepare you for the *executive inquisition*, which you'll eventually face (like the innovator in Figure i.) to justify everything about your idea from its customer value to its financial return on investment (ROI). How you approach this inquisition ultimately decides your fate.

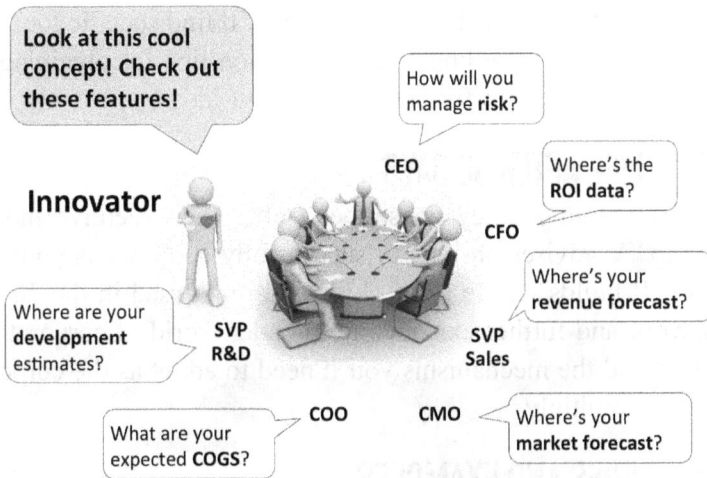

Figure i: An innovator faces the executive inquisition

The Savvy Corporate Innovator is divided into Four Parts. Each part builds off the others to provide a well-rounded perspective. Reading them in sequence offers the most comprehensive detail. Afterwards, you can easily review the key questions and strategies of those chapters you find most relevant to your needs.

PART I: EXPLORING THE INNOVATOR'S WORLD

Innovation is as much about people and your corporation's ethos as it is about ideas. Part I asks you to examine yourself, the nature of your ideas, your corporate environment, and the executive personalities you'll meet along the way.

PART II: A CORPORATE INNOVATION TRANSLATION

The behaviors and expectations of corporate decision-makers often seem foreign to the rest of us. Part II helps you bridge the gaps between you, your ideas, and executives. It also explains how to approach the key challenges of risk, forecasting, and customer value, as well as interprets the executive language you must use when presenting an idea.

PART III: THE 30-DAY ACTION PLAN

To develop a concise, compelling opportunity proposal, you need a plan. Part III synthesizes the most relevant innovation methods into steps that you can execute in 30 days. You'll find specific tools and activities as well as selling and presentation tips for obtaining the first stage of funding to further develop your idea.

PART IV: LIFE AFTER 30 DAYS

Your innovation journey doesn't end after the executive inquisition. Part IV advises on how to successfully continue as your idea evolves. It builds on the same skills and tools used in the 30-Day framework and further explains the broader world of corporate innovation and the mechanisms you'll need to adopt as the corporate challenges multiply.

CASE STUDIES AND EXAMPLES

Throughout this book, you'll also read stories and examples of successes and failures based on real-world innovations. Most of the names and details have been modified to protect the guilty, but not always. In addition, you'll follow two extended case studies that illustrate key concepts and methods. In Parts I and II, "Bob" tries to sell an idea for a new medical device, and in Parts III and IV, "Mark" proposes his roof-cleaning robot idea.

From these case studies, you'll be able to review the material in a (hopefully) fun way, but more importantly, you'll start to relate each scenario to your own experiences, which should encourage you to keep working at the theories and action steps in *The Savvy Corporate Innovator*.

INTRODUCTION

Innovation starts with ideas, but unfortunately, that's often where it ends.

The Problem With Corporate Innovation

Most people realize a company must innovate or die. But innovation success is as much about communication and relationships as it is about technology and products. Problems arise when an innovator and an executive get in a room and try to reach a consensus about next steps. If the innovator talks about product features, attributes, and technology while the executive wants to hear market, revenue, and forecasting data, the conversation won't go well. I've watched many innovators enter a funding meeting with an "idea that will change the world," then leave angry and confused by how quickly it was picked apart in how it might not survive the market.

Many corporations have set innovation processes with plenty of tools and templates. Yet innovators may still not know exactly what to do when asked to present an idea for resources and funding. My goal is to help bridge those gaps, especially if you feel there is an "us versus them" mentality between yourself and the decision-makers who control the funding.

Two Unique Corporate Mindsets

Here's why it can be so hard for innovators and executives to understand each other when it comes to corporate innovation.

An innovator speaks:

"I was hired for my ability to create products. I've had success with my projects and have always delivered pretty much anything the company needs. I enjoy discussions of technol-

---

ogy, features, and design that naturally come with innovation. My problem is that executives are constantly asking me to justify my work, give them a plan, do more research, and explain my design process. As an innovator, I'm focused on developing ideas into real products and find it hard to stop in the middle of what I consider the most critical work just to develop some plan that I know will be wrong anyway. It distracts me from what I believe I was hired to do.

Developing proposals is a perfect example. When I have a good idea, it comes to me with a lot of details—what it looks like, who will use it, how it would work, etc. I often see a full design complete with all the functions. Often I get with my team and get everyone excited. Then we all start talking about features and how to make the product great! I know that market size and sales numbers are important, but I don't have the time or desire to really work on something an executive probably won't even read. That's just not how my brain works."

An executive speaks:

"I'm busy running the details of our business. I have non-stop meetings to review operations plans, solve problems, and hit our numbers. Listening to an innovator talk for an hour about what's likely to be a whim is not always a good use of my time.

The last idea meeting I took started OK. The concept was fine…I got it. But, the innovator went on and on about all the cool features and how no one else could pull it off. I tried to interject with questions about the market and what it could do for our business, but he kept coming back to the technology. I found the whole conversation draining. After 20 minutes I directed him to work with his manager to get the answers I needed.

I need our innovators to realize that ideas are great, but I also need data and analysis to make intelligent decisions. They

Dorian Simpson

must help me understand the market for the idea and its strategic fit with our business. If you can prove your idea has merit, then I'll not only listen, but I'll happily provide you the resources you need and the freedom to pursue it further."

While these two mindsets might seem diametrically opposed, they can—and must—find common ground.

BIG Ideas Are The Starting Point

Ideas are the sparks that become the flame of opportunity and profit. They might start with a simple observation: "I just saw a customer who had this problem. Is there a broader market?"; "I have this technology. What can I do with it?"; or "I just thought of this cool product. Can it make money?" Your ideas might just be earth-shattering and change the world: "Let's create the next virtual 3D social-networking platform that leverages augmented reality, machine-to-machine communication, cloud-computing, and haptics-controlled robotics!"

You know this feeling...

Meet Mark. It's Monday morning and he's sitting in his status meeting trying to pay attention to his manager. However, he can't stop thinking about the moss that's growing on his roof. He spent the weekend trying to clean it off using spray cleaners and brushes. Nothing worked. He imagines the moss nibbling at his shingles and creeping into his bedroom. "Man," he thinks, "there must be a better way." Then the idea strikes! "What if I had a robot that could automatically clean moss from my roof?"

Since Mark is an engineering manager for a robotics company, he's familiar with consumer robotic vacuum cleaners and sees a similar device sliding across roofs around the world. His mind starts racing. "This is a whole new product category for us! We could use our robotic technology and add roof scrubbing components. It would be easy!"

At his desk, Mark sees the pile of work he should be doing, but the image of his new idea won't leave. He gets out his lined paper and starts drawing. "We could even make options to clean moss from sidewalks, walls, and steps! Wait! What about snow removal?" The best part is that he has the technology at his fingertips. What could go wrong? He can't wait to share his idea with the world.

Do you identify with Mark? Most innovators do. New ideas are exciting. But it's at this exact moment that ideas are fragile. How you approach the next step can mean the difference between walking the path to greatness or a hazardous path to oblivion. What do you think Mark should do next? Should he create blueprints for the concept and set up a meeting with his boss to show exactly how it will work? How about a mini-prototype that actually moves? Should he email the CEO? Maybe purchase the URL for MossBeGone.com?

The answer to all of the above is NO. The correct next step is to develop a clear plan for the funding executives who control the resources. This is the step most corporate innovators skip. Whatever your BIG ideas, getting support for the first stage of the project is critical to long-term success.

By the way, we'll return to Mark and his MossBeGone project later. For now, you can imagine him in his shop taking apart his robot vacuum cleaner to figure out how it could gobble up his moss.

The 30-Day Concept Explained

Evaluating ideas, even BIG ideas, must be efficient and timely. Taking 30 days to develop a proposal provides a time limit as much as a time frame. Why 30 days? Any less, the shortcuts could lead to poor results. Any more winds up wasting a lot of time and energy.

The 30-Day concept forces innovators to focus on the needs of the decision-makers. Senior executives always say they want quick turnaround so decisions can account for the increasing speed of the marketplace. Ideas have a shelf life and markets have windows. 30 days is enough time for a savvy innovator to gather the right data, execute critical activities, and develop a clear funding proposal for the next stage of the project. Or you decide to kill it and move on to something else.

Also, don't expect an actual product to be completed in 30 days. A savvy innovator knows that innovation is never linear—you must learn, execute, learn more, and repeat. I'll help you through the steps to move ideas through the first stage of funding. Once you've mastered those, you can use the same skills and methods to move the idea to the next milestone...and then the next! Keep working *The Savvy Corporate Innovator* approach until your BIG idea becomes a new product, a success in the market, a technical acquisition, or whatever other outcome you desire.

You're ready!

PART I

EXPLORING THE INNOVATOR'S WORLD

An Examination Of Yourself, Your Ideas, Corporate Environments, And Executive Personalities

INTRODUCTION TO PART I

Innovation is just another word for realizing
aspirations in complex environments.

In the corporate world, innovation is far more complicated than simply having new ideas. It's a world filled with egos, murky information, and a constant—sometimes confusing—battle over funds. It can truly feel like chaos. Taking an idea from inception to funding in the corporate world should be approached like a journey to another planet full of obstacles and opponents with questionable intentions. The first phase of this journey is developing a compelling opportunity proposal to present at the first funding decision meeting—the executive inquisition.

To prepare for this phase, you must first become aware of what you'll be up against. You must examine yourself, your ideas, your corporate environment, and the executive personalities you'll encounter along the way. Once you understand these elements, it's much easier to succeed as an innovator.

Following is the story of Bob, an innovator, who tries to convince Nancy, a funding executive, to support his BIG idea for a new product. Although Bob is a member of the engineering team, he could just as easily be in marketing, sales, operations, or other functions. Bob's tale will continue after each chapter in Parts I and II so that you can follow his real-life journey to get funding for his BIG idea. Like you, Bob must learn how to navigate the corporate chaos and address all the obstacles he encounters along the way.

BOB AND NANCY:
A Tale of Navigating the World of Innovation

Meet MedCo. MedCo is a company with $720 million in revenue that sells heart-monitoring equipment primarily to hospitals. It has been struggling with declining sales and major competitive pressure on prices and margins.

Meet Bob. Bob is a director of engineering at MedCo. He's confident that remote diagnostics is the future of medicine and believes his plan to add new technology to current products will expand current sales and open up an entirely new market for the company. He's been thinking about his idea for weeks now and has developed the technical architecture and defined some key features. He believes he's ready to share his idea and request resources for the whole project.

Meet Nancy. Nancy is the CEO of MedCo. She's worried about the upcoming board meeting in six weeks. To pick up sales and be competitive again, she needs to show progress and a comprehensive plan on how she'll grow the company and improve both revenue and profit.

It's Monday morning and Nancy has granted Bob a 30-minute meeting to discuss his idea. Bob sits down at the conference table in Nancy's office.

Bob: *Hi Nancy. Thanks for the time today.*

Nancy: *Hey Bob. How are things going? What did you want to see me about?*

Bob: (Handing her a 25-page presentation.) *My team is working on an idea that adds cloud services and wireless technology to our heart-monitoring product. A doctor can monitor a patient's heart from anywhere in the world. We can lead the industry in this new area!*

Nancy: (Ignoring the handout.) *Sounds interesting. Tell me more.*

Bob: *We already have Wi-Fi capabilities, but we don't really use it for anything except when technicians diagnose the equipment. Our heart-monitoring sensors don't support Wi-Fi today because they are hardwired to the machine. But we can add a Wi-Fi chip so they communicate, and we're also looking at Bluetooth. Here's how we can upgrade our current product to connect to a cloud network.*

Bob turns to page nine and shows Nancy a network diagram, including the communication protocol between the heart monitor and sensors.

Nancy understands the concept immediately and tries to interrupt with a question about potential customers, but Bob continues to explain the network and how his new product will utilize wireless technology. Bob seems oblivious to Nancy's disinterest in the details and keeps talking. Nancy finally cuts him off.

Nancy: *OK Bob. I get the idea. What is the market for this?*

WHAT'S GOING ON?

Bob has started to share his BIG idea, but falls into a common trap of focusing too much on technology and product features.

Will he be able to get Nancy's support to fund the project?

(To be continued…)

Chapter 1

EVERY KIND OF INNOVATOR

Various Mindsets And Their Unique Challenges

Your reputation as an innovator will always talk louder than your ideas.

Rate Your Current Skills

- What are you doing to build a reputation as a respected innovator?
- What is your innovation style and how does it impact your success?
- What range of skills are necessary to innovate and what are your own strengths and weaknesses?
- How do you motivate allies to help you develop your ideas?
- How objectively do you view your ideas?

Introduction

Innovators are willing to put ideas out there to be judged by their corporation and even the world with intense and not always positive scrutiny. Savvy innovators use tactical skills to turn ideas into valuable products, technology, and market opportunities. They have mental resilience and strong interpersonal skills to sell ideas to the corporate decision-makers who control the resources. Being savvy means understanding your situation, learning the skills to inspire others, and convincing executives to support funding. A savvy innovator ultimately builds a positive reputation as a respected innovation professional.

As An Innovator—Who Are You?

You're someone who wants to take a BIG idea for a new opportunity and move it forward in your company. You understand the steps and methods to systematically shape, investigate, prototype, challenge, build, and sell an idea until it grows into a real business opportunity. You may be an engineer or research professional tasked to devise product and technology ideas. You might be a product manager who must come up with new directions for current product lines. You may already be assigned to an innovation team, or you may even be leading a team that is investigating a new opportunity. You might also be a customer service agent, sales manager, or finance person who's not responsible for new ideas, but you have them and want to be heard! Innovators come from different fields, but have something in common: you have BIG ideas and need the skills to evaluate, develop, and sell them to corporate decision-makers.

INNOVATORS ARE EVERYWHERE

Steve was a regional sales rep for Sealed Air. The company made Bubble Wrap, polyethylene-foam mailers, and 100% recycled content paper. But customers were always looking for something cheaper and easier. One day, a customer threw

---

a box at Steve. It was loaded with latex balloons used to protect the items inside. The customer said, "If you want my business, come up with something like this where the air won't leak out." On his way home, Steve pulled over and sketched the concept for inflatable airbags for packaging.

Steve's idea started low on the priority list. He was quickly frustrated that his idea wasn't being executed. He thought, "This has huge potential and it feels like we're just sitting on it." To keep it going, Steve had lots of informal conversations where he continued to sell executives on how the idea could change the company. Finally, after 15 months building relationships and persevering, Sealed Air rolled out the product, and Steve became the inventor of VoidPak™ the inflatable airbag for packaging.

Corporate Innovators Are Not Entrepreneurs

True startups, the realm of entrepreneurs, have few constraints in how they navigate markets, build brands, hire people, develop products, and grow the business. In contrast, corporations have these constraints and more. Ideas must match the overall strategy of the company, reflect corporate brands, conform to financial metrics, meet revenue goals, and satisfy the personal and professional needs of executives and shareholders who have their own ideas for the company. A savvy innovator must first and always consider the goals of the company when presenting ideas.

One possible outcome of corporate innovation is taking an idea and incubating it as if it were a startup company within the corporation. This model is often called *intrapreneurism*. The innovator that leads the effort is an intrapreneur. While this is certainly one possible outcome of corporate innovation, and one we'll address later, it's not the only or even the most likely outcome.

Is Corporate Innovation A Full-Time Job?

Developing ideas is rewarding, exhilarating, challenging, and yes, time consuming. Chances are you have a full-time job in one department and must find time to develop new ideas on the side or with an innovation team. Google, for instance, has a 20-percent rule, which gives innovators one day a week to work on pet projects. 3M is famous for its 15-percent rule. These are nice programs, but most companies don't have the luxury of giving spare time for innovation. Usually innovators get no relief during an initial idea-evaluation period. That's life! If you're in a busy role and you really care about the idea, you'll need to find the time to devote to the effort.

From my experience, it takes about 80–100 hours to develop a concise, compelling opportunity proposal. How you fit this time into your job is up to your situation, and in Part III you'll learn how to use your valuable time in the most focused way possible.

Building A Reputation As A Savvy Innovator

While it's nice to think every idea is judged by its own merits, it's almost impossible to separate an idea from its innovator. Your reputation as a savvy innovator will often determine an idea's fate even before you share it. I've seen reputations work both negatively and positively. Often innovators tell me new ideas, and even though I try not to be biased, I find myself judging what they say based on past experiences with them. If they have pitched confusing or inane ideas before, I automatically assume this one will be the same. But, if the person has built a professional reputation, I'm more pre-disposed to like the idea and trust she has done her homework and will fulfill her promises.

Four Types Of Corporate Innovators

Innovators come in all ages, genders, functions, and levels of experience. But long-term success as a savvy innovator is based on two main factors: 1) innovation execution skills, and 2) innovation political skills. Execution skills are the tactical methods that drive projects and execute necessary actions. Political skills help you understand and negotiate the inevitable office politics you'll encounter along the way. These two dimensions create four types of innovators. Figure 1.1 shows how.

Being aware of your own style and the style of others will help you leverage your strengths and mitigate your weaknesses.

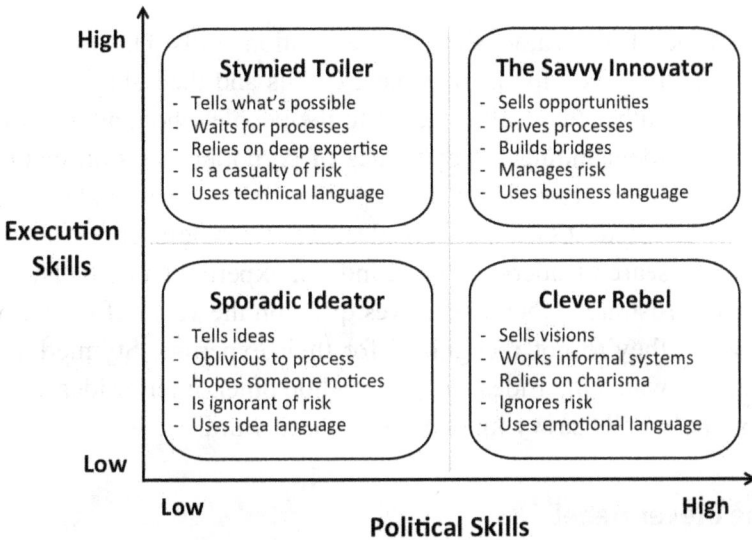

Figure 1.1: Four types of innovators

The Sporadic Ideator

This type of innovator always has new ideas. Unfortunately, they rank low on both execution skills and political skills. They may be engineers, marketers, customer service agents, or sales people who are natural idea generators and see ideas from many sources. They throw out random "idea darts" from all directions with enthusiastic comments like: "We should do this!" Or, "Why aren't we building one of these?" The problem is, they don't know what to do with their abundant ideas. Without a clear path to share and execute their idea darts, these innovators are easily frustrated by the lack of action in the corporation. Sporadic Ideators share ideas with passion, usually with peers, but quickly give up when the going gets tough. When they do get the opportunity to voice their ideas, they usually lack the political savviness to effectively sell them up the chain. This can lead to defensiveness. Sporadic Ideators often believe they have better ideas than most, but no one really listens.

The Stymied Toiler

This type of innovator is high on execution skills, but low on political skills. They focus on project details and their subject-matter expertise rather than selling ideas to the people who matter. By not clearly understanding the dynamics of corporate decision-making, they keep their heads down and hope the company's formal processes will bring success. Stymied Toilers are often engineers or technology or research leaders who are, indeed, experts in a field. They're usually frustrated when executives question the value of their work or when they're not recognized for their expertise. Stymied Toilers often watch their ideas languish while other, inferior ideas, from more politically savvy innovators move forward.

The Clever Rebel

This type of innovator ranks high on political skills, but has low execution skills. They rely on gut feel and interpersonal skills to move ideas forward. Clever Rebels are savvy enough to inspire peers and enlist their help to relentlessly pursue ideas. They don't get discouraged easily and may even risk personal or corporate success to move their ideas forward. Clever Rebels don't like processes because they just slow them down. If they have any formal power or resources at their disposal, their ideas often become pet projects that operate "under the radar" unless killed by a more powerful adversary. Clever Rebels also usually lack self-awareness about their innovation skills and are prone to ignore warning signs in the market or technology. They often get frustrated by unsuccessful progress, which can cause them to leave the company for more hospitable pastures.

The Savvy Innovator

These innovators have a balanced blend of execution skills and political savviness. They are self-aware, appreciate their strengths, and know when to ask for help. They deeply understand the stages of innovation and the importance of mitigating both technical and commercial risk. Savvy Innovators know that ideas are merely starting points that can turn into a variety of opportunities. When selling their ideas, they build bridges between themselves, other innovators, and senior executives. They're always developing as they continuously learn both the execution and political sides of innovation. Savvy Innovators also understand and use existing processes and address executives on a personal level when asking for resources. Most importantly, Savvy Innovators are not overly attached to their ideas. They know when an idea is not right for their company.

I've worked with all these types and watched with fascination how each one sells their ideas. Take Keri—a Clever Rebel who was a manager at a test equipment company and a master of gaining support for her ideas. Keri could weave a great story that inspired peers and executives alike. You couldn't help but leave her office excited to help her see her idea come to life. She could get projects started, but they often fizzled out unless managers who could execute better took them over. Sometimes her ideas were successful. Other times they just lead to frustration.

Then there was John. He was a product manager at a media technology company—a classic Stymied Toiler. He had complex ideas that always came with detailed presentations, technical diagrams, and pages of "great data." Clearly a deep thinker, John couldn't sell an idea if his life depended on it. Time and again his audience's eyes would glaze over after 15 minutes of mind-numbing details. He rarely received the go-ahead for an idea unless a more politically savvy person helped him.

To become a Savvy Innovator you can certainly start as one type and improve your skills in weak areas. Depending on the opportunities, different innovator types can thrive in different environments. First recognize yourself. Then think about how you operate and how you can learn from other styles.

Six Essential Innovation Skills

Developing an idea even to the point of asking for first-stage funding requires a wide range of skills from technical abilities to critical market analysis. To become a savvy innovator you need execution skills in the following areas of expertise:

Data Gathering: finding and interpreting relevant data.

Market Analysis: defining a market and estimating customer demand.

Customer Insight: understanding and validating customer needs.

Technology Exploration: applying technology to solve customer problems.

Go-to-Market Tactics: developing preliminary tactics for sales, marketing, manufacturing, and other execution challenges.

Financial Scrutiny: estimating revenue, costs, and expenses to the satisfaction of decision-makers.

This doesn't mean you must have all of these skills yourself. We'll address many of these skills in this book, but few innovators inherently have a complete skill set. Technologists are often weak in marketing while marketers may be weak in technology. People spend years learning the skills needed in a specific field, but innovation requires that you broaden your skills to the point you can at least understand the basics with enough awareness to evaluate your current strengths. It's important to know your weaknesses and look for help in the areas where you need more expertise.

Leveraging Team Members

Developing ideas at the early stage can be a lone innovator's task. But because an innovator needs help with a wide range of skills, there's a big advantage to working in teams building a well-rounded opportunity proposal. Figure 1.2 shows how a marketing expert brings knowledge about markets and what's important to customers.

The technical person brings expertise on what is possible based on internal company technology, technical trends, and access to external technology. It's extremely rare to find those skills in the same person.

Great success often just requires two people—a market expert and a technical expert. That's pretty good for a minimum team. Just look at Steve Jobs/Steve Wozniak or David Packard/William Hewlett. I once worked with a Seattle company that had a proposal contest where they formed teams to identify new opportunities and present their ideas to the executive team. Out of eight teams that pitched their idea, the team that won had just two members—a technical expert and a savvy business-development leader.

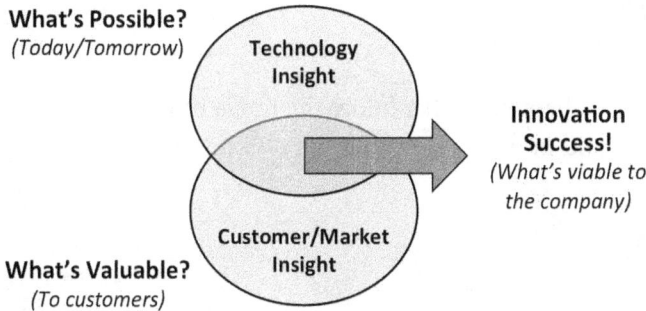

What's Possible?
(Today/Tomorrow) Technology Insight

Innovation Success!
(What's viable to the company)

Customer/Market Insight
What's Valuable?
(To customers)

Figure 1.2: Teams can aid success

SAVVY IDEA: AVOID IDEA BLINDNESS

We love our ideas, but we can easily run the risk of idea blindness. This happens when we're so close to an idea that we ignore contrary information. We may even feel offended when asked reasonable questions as we start the funding process. I fell into this trap many times early in my career. I once created a company called Hobby Starters. It was at the beginning of the e-commerce era, and I was sure that an online portal for hobby-minded people would be the next big thing. I conducted a range of interviews but fell into the trap of idea blindness. When people told me that price was very important, I heard, "Price could be higher if I offered something truly unique." When someone asked, "Can't I get all of these products at craft stores?" I told myself I could overcome this with a better customer experience.

I proceeded to build the business, complete with distribution agreements and a full website, only to find out I hadn't sufficiently addressed the customers' concerns. I shut the site down after six months and had to rethink everything. I learned from this experience. It's crucial to be as objective as possible and avoid idea blindness even if you are sure your idea will be a huge success.

In the next chapter, we'll explore the range of ideas that corporations generate and how to determine which ones are worthy of pursuing further.

CHAPTER 1 REVIEW—EVERY KIND OF INNOVATOR

Innovators come in all types and can be found in every function. A savvy innovator knows his strengths and weaknesses, has the ability to execute the tactics of innovation, and knows how to navigate tough political environments to sell ideas. Building innovation skills takes time, but becomes more natural as you train yourself to recognize and react appropriately in all situations.

Five Savvy Success Strategies

As you hone your skills, practice these five strategies to start on your path towards becoming a savvy innovator.

1. **Never assume your ideas will sell themselves.** It's rare the others will see the same potential that you do. Ideas must be marketed and sold as much as any other product or service out there. The "market" for your idea is your company and the customers are corporate executives.

2. **Ensure you've embraced a corporate mentality.** Corporate innovation is not about the best ideas; it's about the best ideas for the company. Corporations must manage risk. Not only do you need to understand corporate risk, but you must understand what's at risk for the peers and executives you need to support your idea.

3. **Learn to love data as much as your ideas.** While most executives love a new idea, it will never be sold on anecdotal information. Quantifiable data is the realm of decision-makers. The more relevant data you have to support why people should risk their job for your idea, the easier it will be to get support.

4. **Take an honest assessment of your own skills.** No one person will likely have all of the innovation skills necessary to develop an idea. Take an honest assessment of your strengths and weaknesses when attempting to evaluate and sell your ideas.

5. **Acknowledge you can't innovate alone.** Even if you have all of the skills necessary to be a savvy innovator, you still need allies. These allies might be functional experts, peers, or executives. To execute any idea worth pursuing, you'll need help along your innovation journey from the moment you start sharing your ideas.

BOB AND NANCY (CONTINUED)
Bob, a classic innovator:

Bob had started sharing his idea for a new wireless heart monitor. Nancy, his CEO, had just asked, "What is the market for this?"

Bob: *Great question. This is perfect for people having their heart monitored. At home, in hospitals, nursing homes and even just walking around! Here is the app a doctor would use to access the equipment. I didn't have time to do a full working app, but here is what I have so far.* (He pulls up a crude screen on his tablet and starts to login.)

Nancy: *That's OK, Bob. I don't need to see the app yet. Who will actually buy this? What do you think the market is for this?*

Bob: (Isn't she listening to me? I just told her! Can't she see how great this is for the company?) *If I were a doctor, I'd demand this product. Along with the heart monitoring, we can also add other functions for monitoring glucose and just add sensors as it expands.*

Nancy: (He's ignoring my questions. I don't have time for this.) *OK. I think I get it.*

Bob: (Yes! She's approving the project.) *Great. I'd like to start writing the product requirements so I can get this into our development process. I'm actually almost done and have started scoping out the resources we would need.*

Nancy: *Hold on Bob. That's great you've put a lot of thought into this, but let's talk through this some more. I still have a lot of questions.*

WHAT'S GOING ON?

Bob is a typical Stymied Toiler, but he has the aspiration and ability to become a Savvy Innovator. He spent a lot of time thinking about his idea and assumed he could get a quick approval from Nancy. However, Nancy wasn't close to committing funds. She saw a lot of potential in the idea, but wondered why Bob came to her so unprepared to answer her questions.

Can Bob get on track and get the guidance he needs to move his idea forward?

(To be continued...)

Chapter 2

THE NATURE OF IDEAS

Types Of Ideas And Those Worthy Of Time And Resources

There are ideas and then there are ideas worthy of 30 days.

Rate Your Current Skills:

- How quickly can you determine whether pursuing an idea is worth your reputation?

- What types of ideas do executives most want to hear about?

- What are the criteria that separate good ideas from the right ideas?

- How do you prioritize your ideas and ensure you're working on the right ones?

- What types of ideas require proposals?

Introduction

Everything accomplished in the world starts with an idea. Ideas can spark revolutions or be the seeds of great companies. They might just trigger what you have for lunch. We must think deeply about our ideas so they can have the most positive impact to our career, our company, and our world.

Innovators and companies typically waste too much time and far too many resources on pursuing the wrong ideas. You have a short window and there are only so many resources available. A savvy innovator separates average ideas from *30-Day Ideas*, the ideas that will be the most successful for yourself, your corporation, and your executives. These are the BIG ideas worthy of spending 30 days to develop further.

Great Ideas Are Not Always The Right Ideas

While it's tempting to run with an idea because it's exciting, feels right, or has huge potential, it's essential to remember that we're talking about corporate innovation. Corporate ideas have to pass more rigorous tests than startup ideas. Because there are so many possible directions a company can take, a savvy innovator must closely examine an idea to determine whether it's worth pursuing one day, much less 30 days, of valuable personal and company time.

Your company is inundated with ideas from customers, managers, suppliers, and other sources. Some are big. Some are small. They're all inherently fuzzy. They range from, "You should change the color on your website," to "You should develop this radical technology." You may be tasked with developing your own ideas or asked to evaluate ideas that originate from other sources. But what exactly is an *idea?* To answer that, let's explore the four types of ideas that are usually generated in the corporate world.

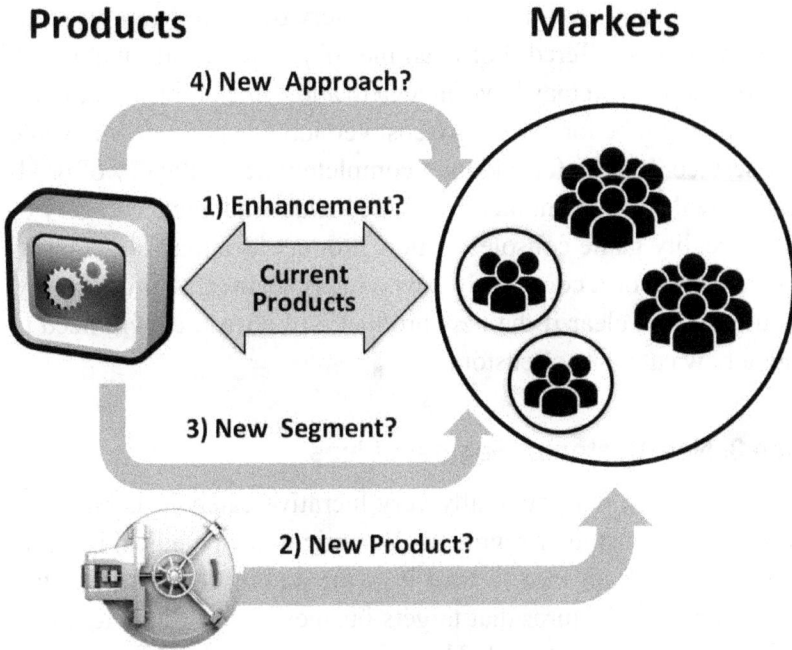

Figure 2.1: Four types of ideas

Type 1: Product Enhancement Ideas

These are the most common and are usually based on adding or improving features to existing products or services. Maybe adding a safety feature to a commercial band saw or voice control to an audio/video receiver. Maybe accessorizing a product, improving a user interface, or creating unique packaging. Sometimes ideas come directly from customers who want to see an improved product. Sometimes they originate from product managers and technologists who are planning future variations as part of their product roadmaps. In most cases, these ideas focus on improving current products to increase sales or stay competitive.

Type 2: New Products and Service Ideas

This type of idea considers something very different from what has been previously offered. For example, if you're already making IP video cameras, you may have an idea to add a line of infrared cameras. Ideas might be for a new low-cost version of a product or a whole new service. Maybe for adding a completely new category of products, as with a TV manufacturer that considers adding a category of virtual reality game consoles. A new product idea may target existing customers or a completely new market segment. Unfortunately, it's usually not clear if the new product will attract, or will need to find, a new category of customers.

Type 3: New Customer Segment Ideas

A less common, but potentially very lucrative category is based on targeting new customer segments. You might have medical imaging technology with an idea to take it to the automotive market. Or a sports-cup manufacturer that targets businesses but wants to create a direct-to-consumer brand. These ideas often start with questions like: "What if we created a coffee drink specifically for athletes?" or "The Latino market is growing rapidly. What are we doing for this market segment?" Usually they come down from senior management, the sales team, or internal market analysts and industry reports.

Type 4: New Approach Ideas

This last category doesn't fit neatly into the others but often includes the most innovative ideas. Maybe it's for changing the business models, such as, "How can we convert our hardware business into a services-model business?" Maybe it's for developing or implementing new technologies, which then ultimately lead to new products or applications. For example, you might have an idea for a new wireless-communication protocol or higher-density memory chips. Or maybe you're a soft drink company that wants to create a new interactive game to sell more cola. New approaches can also be operational processes and improvements, such as creating an idea-management system or finding ways to outsource manufacturing.

Ideas Range From Incremental To Radical

Another way to categorize ideas is how far they veer from current business practices, ranging from incremental to radical. An incremental concept idea is similar to the types of products you develop and customers you serve today. They have lower risk and can usually be quantified with available data. They fit nicely on current product roadmaps and can be integrated easily into operations plans. They may not even need a big proposal for funding or executive support. Radical ideas come with higher risk and are much more difficult to quantify. They force changes to the status quo and cause people to seriously debate whether they are right for the company. Radical ideas almost always require a solid opportunity proposal and all of your political and execution skills.

A savvy innovator knows the further an idea is from the core business, the more difficult it is to get support from the organization. While radical ideas come with the most risk, they often bring the greatest reward. You need to be highly skilled and knowledgeable to move these ideas forward.

THE STARBUCKS EXERCISE

Innovators and companies face thousands of ideas in all these different categories and more. Imagine the sorts of ideas a company like Starbucks might innovate.

Product enhancements:
- Changing the formulation and adding/subtracting other ingredients to make the drink less bitter, more caffeinated, sweeter, more natural, etc.
- Creating things on the cup such as adding sponsors or

applying scratch-and-sniff features or designing cups by well-known artists and musicians.

- Improving the cup (and/or top) so people don't spill on their clothes. (I always have to turn the lid opening away from the seam or it leaks on my shirt!)

New products:

- Adding related products: candy, ice cream, clothes, beans, presses, grinders, baby apparel, etc.
- Creating coffee-based games or other mobile apps.
- Adding business services, movie kiosks, etc. to the 10,000 plus Starbucks locations.

New market segments:

- Segmenting the market and identifying new products or experiences to better target business users, moms, health-conscious adults, socialites or teenagers.
- Considering mobile coffee carts or new ways to reach different types of customers based on lifestyle and activities such as biking, soccer, etc.
- Entering new countries or acquiring new types of beverage stores such as those selling smoothies.

New approaches:

- Creating a new in-store experience using robotic dispensers that give customers an unlimited ability to make lattes precisely how they want them.
- Using technology to detect when customers walk into the store so that their same, prepaid beverage is ready.
- Offering a subscription service like Netflix for coffee. Pay $24.99/month and get whatever you want eight times a month.

The list goes on. I'm sure Starbucks has thought of these and many others from the most mundane, "Let's greet everyone when they come in the door", to the most wacky, "Let's create 'cones of silence' where people can make phone calls in privacy without annoying others."

Get the idea?

Unfortunately, Ideas Don't Come With Labels

The ideas in these categories are just the initial sparks of opportunity. Categorizing them can get fuzzy since one idea always leads to the next. And then the new idea often shows up in a different category. For example, an idea to target a new market segment will ultimately generate product ideas. A new business-model idea will naturally force ideas about new market segments.

What category would Apple's iPad have been in when the innovator recommended it? It's certainly a new product, but some might argue it's a new market segment or a product enhancement. Since the iPod Touch and iPhone were already in the market, was the iPad just an incremental product enhancement? The concept was certainly not radical…or was it? Think about how the iPad created a radical breakthrough that exploded a nascent tablet computer market into a massive industry. Categorizing ideas can help frame our thinking, but since they don't come with labels, it's up to you to determine how to explain the idea.

Most ideas needing significant investment will require some form of business justification, but not everything needs a formal opportunity proposal. If it's a simple idea such as starting a company blog, then a simple recommendation that takes 30 minutes to outline key costs and advantages might do the trick to move the idea forward. Incremental ideas like product enhancements or operations improvements may be valuable, but you can usually handle them by a standard process of prioritizing resources during product or operations reviews.

The problem is ideas can go on forever. At some point a savvy innovator must stop the ideation process and ask tough questions, "Is this a good fit for me and my company, and does it justify being evaluated further?" Not every idea is a 30-Day Idea.

30-Day Ideas Are Special

A 30-Day Idea is bigger than your average idea. Usually it hasn't been included in the current budget or operating plans, but it will also compete with every other unplanned opportunity that requires funds. A 30-Day Idea makes an executive say, "We need to really think about this. Send me a proposal." The company might need to rethink its current skills, channels, personnel, or strategy. Anything that affects sales quotas, impacts an executive's favorite product, or threatens a VIP is also probably a 30-Day Idea.

In other words, an idea that forces the company to veer from its current path and requires serious justification is the first filter for a 30-Day Idea. To get the resources you need to execute, you'll need more information, sound financial judgment, and an opportunity proposal that meets the needs of your company and its decision-making executives.

But Not All 30-Day Ideas Are Equal

How do you know whether it's worth spending 30 days analyzing the value of the idea without actually developing the idea? This decision is especially difficult when you've got many ideas that seem worthy of analysis. Later we'll address this in more detail, but for now, see if your idea passes the 30-Day Idea Five-Point Inspection.

- **Is it aligned with the company's vision, strategy and goals?** To get any traction in the corporate setting, it must be headed in the same direction that executives want to move.

- **Is it big enough to be valuable to decision-makers?** It may generate $5 million, but for a company that makes billions in revenue, it won't even raise an eyebrow.

- **Is there a reasonable path to execution?** If there's a large gap between existing operations, such as R&D or marketing expertise, compared to what is needed to execute your idea, there is probably too much risk

- **Is there a reasonable path to test the opportunity to mini-mize risk?** Since corporate innovation is all about risk, there must be a means to test all aspects of the concept or it will rarely get approval.

- **Are you passionate about pursuing the idea?** Innovation takes energy, resilience, confidence, and persistence. Having a certain amount of drive will be necessary to weather the corporate storm.

This isn't a black-and-white checklist. It's OK if you answer "maybe." But it's essential to make sure an idea has potential merit from the beginning. If even one answer is a clear 'NO', you'll need to take a serious and objective look at the idea. Don't waste time just brainstorming great ideas. See if they have real potential for the company and go from there. 30-Day Ideas are powerful! They're good for you, your executives, the market, and your corporation as a whole!

30-Day Ideas Need To Be Prioritized

It's imperative to quickly evaluate and prioritize ideas because even 30 days is lot of time. So try a simple scorecard using the 30-Day Idea Five-Point Inspection. For example, assign a rating to each criterion on a scale of 1 to 5. (5 means the idea is perfectly aligned with corporate vision, strategy and goals; 1 means it's not close to being aligned.) Once all of the ideas have been rated and ranked, select the top-rated ones for further evaluation.

The secret to using a scorecard and getting executives to support your ideas is making them part of this process. This includes both developing the criteria and ranking the ideas. If executives haven't participated in the selection process, you risk them not agreeing with you later and questioning your results with: "Why did you use this criteria?" and "What makes you think this rates a 4?" A savvy innovator knows how important it is to have active participation by decision-makers in the development of the idea.

The Idea End-Game Will Vary

Once a 30-Day Idea has been selected, your goal is to develop an opportunity proposal that asks for the first stage of funding to move it forward. This doesn't mean you'll suddenly get a big budget to form development teams or a new business unit. In fact, those outcomes rarely occur. When an idea is being evaluated, anything can happen. Outcomes can vary. You won't always know what that outcome will be when you start developing an idea. It might be to merge the idea into operations plans, make it the basis for another project, spark a partnership or acquisition, or get morphed into other ideas.

Even if your idea doesn't move forward, killing it should still be considered a successful outcome. You'll know you've done your best to grow the idea into something valuable, but it just wasn't meant to be. As one innovator said, "I wish I'd known how to really challenge my business concept earlier so I didn't waste three years of my life on it."

SAVVY IDEA: USE PROBLEM STATEMENTS TO GENERATE GREAT IDEAS

When a customer identifies a big problem, you have a valuable starting point for innovation. Take advantage of a customer's situation and use a problem statement to start investigating questions such as: "Shorter customers hurt their backs when they sit too long in our chairs. What can we do about it?" Or, "Some of our vegetable processing plants are worried about contamination. How might we help them identify more biological hazards?" And another, "Our customers' biggest complaint about our contact lenses is dry eyes. Shouldn't we dedicate innovation time to this?"

Problem statements don't need a solution in mind to start an investigation. In fact, it's often better not to have one in mind so your thinking won't be biased.

Dorian Simpson

In the next chapter, we'll explore the variety of corporate opportunity environments you will encounter and the challenge of navigating each one.

CHAPTER 2 REVIEW—THE NATURE OF IDEAS

Not all ideas are worthy of your valuable time. Ensuring an idea is a 30-Day Idea before moving forward saves time and builds your reputation as a savvy innovator. An idea is right if it has value to you, the company and the company's executives. Once you learn to not just generate ideas, but propose the right idea, you'll gain more and more influence as a leader in the company.

Five Savvy Success Strategies

As you hone your skills, practice these five strategies when making the decision to move forward with your next big idea.

1. **Marry ideas to customer needs, not personal desires.** We love our ideas! It's easy to assume customers and executives will love them too. Focus on solving a real customer need that can be proven valuable. Separate ego and opinion from the objective evaluation of ideas.

2. **Get input early to determine if your idea is a 30-Day Idea.** Don't waste a lot of time on ideas that have no chance of getting funding. Seek a trusted executive to check the idea against the five-point inspection and uncover the most problematic points.

3. **Think of your personal time as a valuable resource.** If you're not passionate about your ideas, it will be difficult to weather the corporate challenges. You don't want to be overcommitted, but you should be serious enough about the idea to dedicate 30 days towards building an opportunity proposal.

4. **Scale your ideas.** While you don't want to waste time on an idea that doesn't make sense, don't abandon it too quickly just because it didn't pass initial inspection. Always ask the question, "How could I expand or modify this idea to warrant further investigation?"

5. **Be flexible on the outcome.** Going through the process of evaluating an idea will lead to many possible outcomes. Perhaps your idea will quickly morph into something else. Change your perspective. If you're in the business to help your company, remember the big picture—success means company success. If your idea is not used exactly how you imagined it, maybe there's another path.

BOB AND NANCY (CONTINUED)
Bob goes deep into his BIG idea.

Bob isn't sure now where he stands. He thinks Nancy likes his idea, but she seemed to want information that he doesn't have. She simply said, "I still have a lot of questions."

Bob: (My requirements document should so the trick) *Let me send you the requirements document I started. That should answer a lot of your questions.*

Nancy: (She sighs, oh great. More technical details.) *OK. This idea does sound like it has potential. How close are the new wireless sensors to what we do today?*

Bob: (He turns to page 15 of his presentation that shows the preliminary development schedule.) *We have a lot of the technology, but need to develop a new sensor with Wi-Fi capabilities. We also need to develop a new battery. The current batteries on the market won't last long enough unless we develop a charging unit. But I doubt that doctors would want to take time recharging the sensors every week. We may not even use Wi-Fi since it's energy consuming. That's one of the unanswered questions.*

Nancy: (hmm...sounds like he doesn't even have a handle on the technology. Who knows when we'll see real revenue from this product?) *So it sounds like there are still a lot of unknowns here. You probably know that our innovation budget is very limited. What makes you think we can work through all of this?*

Bob: (Man, she doesn't trust the development team. He turns to page 18 that shows a detailed Gantt chart and the number $3.5 million at the bottom.) *It's all doable. It'll just take some time. Yes, there is risk in the schedule, but I'm confident we can hit it. We just need to work through a couple of technical* *hurdles. We may even be able to accelerate the schedule. I think we can develop the product for about $3.5 million.*

Nancy: ($3.5 million? I'm sure he hasn't considered all the marketing and production expenses this will take, and I've never seen a schedule that has been hit before.) *I think your idea has potential, but let me ask you a couple more questions.*

WHAT'S GOING ON?

Bob believes he has a great idea, but is it a 30-Day Idea? Nancy likes it, but she's not sure it has significant revenue potential or that the development team has a handle on the technology.

Because he's blinded by his passion, he doesn't seem worried about the $3.5 million investment (after all, it's not his money). But to Nancy, Bob simply has not been able to identify a clear path to lower the risks associated with the concept.

Instead of spending months working on the technical details and product description, Bob needs to focus now on determining if this is the right type of idea for MedCo before he begins a detailed analysis.

Can Bob get beyond the innovation and see his concept from a corporate point of view?

(To be continued...)

Chapter 3

CORPORATE OPPORTUNITY ENVIRONMENTS

Identifying And Navigating Your Own

Corporate organisms don't mean to bite; they are just protecting themselves.

Rate Your Current Skills

- What exactly are your company's strategic goals and major initiatives?

- How well do you understand the real approach your company uses to select new opportunities (besides documented processes)?

- Where does the power really exist in your company for making key decisions?

- What resources are available to you and how do you gain access to them when evaluating a new idea?

- How accurately can you describe the type of innovation environment your company has created?

Introduction

Corporations are complex entities. Even with formal innovation processes, a company's culture, dynamics, and historical experiences usually determine an innovator's success when seeking funding for a new idea. Each company has its own "personality" that has evolved over many years. I once worked with an Oregon lumber company that started in 1850. You could still *feel* the founder's presence in everything the company was doing to that day.

Every innovator faces pros and cons within the confines of a company. But regardless of the environment, there's always competition for new ideas. A savvy innovator understands the corporate environment and has the skills to optimize the positives and work around the challenges to prosper and get the resources for ideas.

Corporate Dynamics 101

Whether public or private, the nature of a corporation is to get a large number of people acting as a single entity for a common purpose – make money for its owners. Under layers of mission statements and strategic plans is this fact: Every individual, from the CEO on down, acts according to his or her personal beliefs and attempts to satisfy personal needs. There's a natural conflict. One person's needs may be at odds with another's. This is what we refer to as corporate politics. You should expect politics to be as normal in your job as getting a paycheck. Your goal as a savvy innovator is to learn the dynamics, rationale, and personalities in your environment and ensure they work in your favor.

ONE STORY OF CORPORATE POLITICS

Rick was the VP of Business Development for a consumer electronics company. It had been extremely successful with analog cable equipment, but Rick was concerned. He saw that digital technology was getting a lot of attention in the industry. His engineering team agreed that digital was the

---

future. They even had a prototype for a new cable set top box that would "revolutionize the industry." Rick's manager, the president of his division, saw it too. She told him one day, "It's time to get this out of the lab and make a business out of it. Make it happen!"

Rick then built a proposal for a new digital product line and took it to the CEO and board of directors for funding. However, they had a culture where risk was to be avoided at all costs. Rick's division was just one small part of the overall company, while another division, that sold analog products, was the company's main source of revenue. They looked at the digital technology and said, "That's no big deal. Digital will never be a real business and customers won't pay for this." Despite the entrepreneurial spirit of Rick's division, there was a gorilla in the way.

The large division influenced the board to move slowly and make sure not to cannibalize their highly profitable business. Thanks in part to the CEO who also saw the digital trend, the board finally funded the project, but not to the level Rick's division needed to commercialize the technology. Rick eventually landed a deal with a key customer for over a million units. The large order finally convinced the board to make a major investment. "It was like a 3-ring circus," Rick said later. "If we didn't get that order, I think the whole company would have gone down with the analog sinking ship."

Corporate Organisms Make Idea Immunity Natural

A savvy innovator must accept that corporations are naturally immune to new ideas. This shouldn't be surprising. Every company survives and gains success by developing certain competencies: hiring people with specific skills, building well-known brands, and becoming experts on specific customers, markets, and technology. Companies even apply lean thinking to every operation to be even more efficient and purposeful toward dominating the current environment.

This is all good—except when the organism senses an invader. Ideas that challenge the status quo feel like an attack. Antibodies, in the form of people and systems, show up to ignore the invading idea, hoping it goes away. Executives or managers might verbally squash it with statements like: "That is not what we do." Or, "We can't do that." Process hurdles might be placed in the invading idea's path with commands like, "Submit all ideas with a detailed analysis that include five years of cost and revenue projections."

The company organism is naturally trying to survive by removing the risk of the attacking idea. This immune system builds up over time based on a variety of factors:

Balance of power. Potentially thousands of people in sales, marketing, R&D, operations, and upper management are all focused on near-term revenue and profits. They're motivated with objectives, salaries, bonuses, and retirement plans based on maintaining the status quo. Conversely, there's a small band of innovators focused on new opportunities for future profit. In a head-to-head battle, say during a corporate planning meeting, usually the masses with more resources, authority, and power wins.

Processes not tuned to innovate. Sometimes companies say they innovate as part of their product development or lean improvement processes. These processes are great for managing incremental improvement projects, but are horrible about managing innovation efforts. If a big idea comes up, it's added to the list of current projects. Unfortunately with so many discussions about next-generation products, sales targets, and operations issues, new ideas get very little attention.

Corporate competencies. Every MBA student learns that a company must leverage its core competencies; from technical skills to customer relationships. It's hard to argue not to take advantage of your strengths, but the world is constantly changing. New ideas often require a company to build new competencies.

Unfortunately addressing short term needs usually takes priority over building new skills.

Budgeting. Every annual budgeting process starts with a discussion of last year's budget then moves on to create next year's budget. There is rarely a budget line item for "unforeseen opportunities." If an idea shows up after budgeting time, funds might not be available to explore an idea until the next budgeting period. By then, the opportunity is gone or the innovator gives up and moves on.

Other corporate constraints. Besides the budget, there's a range of other constraints against innovation: revenue targets, corporate brands, and margin needs. Even strategic vision can limit the types of ideas that get heard.

Consider how idea immunity can affect a company's success. In a slow moving market that doesn't force change, idea immunity may take years or decades to affect success. In a fast moving, competitive market, idea immunity can have a dramatic impact in a short period of time. Look at Motorola—quickly impacted by its immunity to app-based smart phones. It was focused on the highly successful Razr product line when true smart phones hit the market. The results were disastrous and forced Motorola to sell its mobile phone unit to Google.

These factors will certainly create challenges for you to move ideas through your company. But there are definitely positives to being a corporate innovator.

Advantages For Corporate Innovators

You get to keep your job. Whenever I've worked on a high-risk project (like AT&T's plan to develop the 'seven levels of computing', which I think it killed after hitting four levels), I was always able to keep my job and move on to the next project.

You're surrounded by big brains. People love to share what they know, and there are always subject matter experts available

---

to help if you just reach out. Technical, marketing, operations, and financial gurus are usually just an internal call away.

You have free market information. There are a ton of great resources (sometimes untapped) available to innovators. Corporations often purchase services from major research firms, have access to industry analysts, or have internal market research teams and tools to conduct research if they are properly guided and motivated.

Your access to customers. Most entrepreneurs would kill to have access to the range of customers that are available to corporate innovators for exploring and validating ideas. Accessing these customers is not always easy, but is essential for your success.

You get to play with cool tools. While I was at IBM, I loved doing designs and having the internal machine shop crank them out. Occasionally I'd sneak in a few designs to prototype a fun idea. Not every company has a world-class machine shop, but most big companies have something for you to play with, even if it's just a 3D printer or access to an online survey tool. Use those tools! They're there for a reason.

All of these are huge advantages over trying to innovate outside of a corporation. Experts might cost thousands of dollars a day. Market data can cost more than $10,000 for a single report. And a mentor who will meet with you every week just walking down the hall is priceless. A savvy innovator uses all of these assets while working through the hurdles unique to every environment.

Four Corporate Opportunity Environments

To determine how you might seek funding in various types of companies, it's helpful to understand specific environments for opportunity development more deeply. There are four categories based on two dimensions: 1) Idea Generation, and 2) Opportunity Processing.

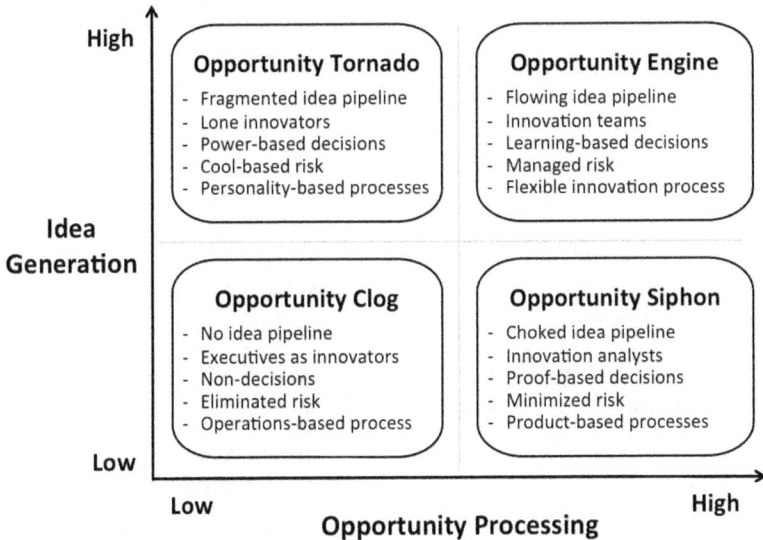

Figure 3.1: Four types of corporate opportunity environments

The idea generation dimension is the spectrum of how well a company captures and generates ideas. It includes the quantity, and more importantly, the quality of ideas. Some companies are high on the spectrum because their industry naturally generates ideas, or because they have solid ideation efforts with programs to stimulate ideas. Too many ideas can be disruptive; too few can stagnant the environment.

The opportunity processing dimension is the spectrum of how well a company manages and processes the number of ideas they have identified. A company that scores high on this spectrum has methods and skilled people to evaluate, select, and drive ideas from initial concept to viable business opportunities. Figure 3.1 shows how these two dimensions create four types of opportunity environments. Let's look at the characteristics of each and how you might navigate each one.

The Opportunity Clog

This environment is low on both the idea generation dimension and the opportunity processing dimension. Few innovative ideas get generated. Those that do often don't make it through to execution. Companies that have opportunity clog environments have dedicated processes for reaching financial goals and improving operations, but not for encouraging innovation. You'll see things like suggestion boxes, but these are used mostly as filters so the CEOs are not inundated with a lot of ideas they really don't want to hear. The clog tries to stay profitable with as little risk as possible.

They're typically in slow-moving markets. Their products, services, and systems have served the company well for years, so why change? They're often run by a small number of controlling executives who rely on experience and gut feel to evaluate selective ideas. Ideas that come from below the senior staff have little chance of getting funding unless the CEO immediately sees an obvious ROI.

As an example, I worked with an industrial pipe company that said they wanted growth opportunities. Management pointed me to its annual operations plan with a list of six ideas such as "developing home-construction products." When I reviewed five years of previous plans for other opportunities they had considered, I found the exact same list each year. To fix this, the company had to learn how to process ideas as well as how to fill their pipeline with a new range of opportunities.

A savvy innovator in an opportunity clog environment must build direct relationships with senior executives and the CEO to understand their concerns and prove the opportunity will succeed. This can be challenging. You must be persistent, do a massive amount of due diligence, and demonstrate success with almost zero risk.

The Opportunity Siphon

This environment is high on the opportunity processing dimension, but low on the idea generation dimension. These companies may have a range of business processes in place such as Six Sigma, 'stage-gate' development, Balanced Scorecard, and others. It's very difficult for them to manage the uncertainty of new ideas. Why? Because opportunity siphons look for perfect ideas with a clear ROI and little risk. Senior management recognizes the need for innovation and may even hold campaigns that lead to ideas. However, when real resource decisions are made, the lowest risk options are selected, leaving most ideas on the floor. Siphons wait until the pressure builds from competitors or key customers before they "siphon" off the top opportunities and actually execute.

Opportunity siphon environments are not always bad. One could argue Apple is an opportunity siphon. For its size, there aren't a lot of dramatically new products coming out. But when it does attack a new category, it meticulously and flawlessly understands the risk to obtain the reward. However, many siphons wait too long to execute opportunities until more nimble competitors have already gained market traction. It's often too little, too late. Intel can be a siphon that struggles with its innovations. From personal electronics, to mobile processors, to TV services, Intel has consistently announced and then killed new products that don't immediately pay off with high returns.

A savvy innovator in an opportunity siphon environment must show fast revenue and profit growth. New opportunities are judged on the same metrics as current products, which can be frustrating for many innovators. To be successful, you must use relevant data at every stage of the innovation process to prove ideas have manageable business risk. Tenacity and professionalism pay off in the siphon environment.

The Opportunity Tornado

Companies that have an opportunity tornado environment love a good idea. You can hear the halls ring with, "That's great! Let's do it!" They're high on the idea generation dimension, but low on the opportunity processing dimension. Tornadoes have an interesting mix of informal processes with strong personalities sprinkled throughout the organization. They seem chaotic with many innovators from all areas of the company (some without any formal authority or budget) trying to move ideas forward. They almost always have documented product development processes in place, but no clear ways to filter, prioritize, or kill new opportunities that haven't already entered the pipeline. There's no rule book to follow, so people end up making their own rules.

Top sales leaders listen to key customers and drive new features and products directly with engineering. Likewise, the R&D group works on its own pet projects for new technology and cool products. When I worked with Motorola when digital TV took off, the possible applications, products and customer requests for our new digital-compression technology were endless. It was an idea-rich environment, but we struggled to manage them all since we had a very low process dimension. We had an opportunity tornado environment where ideas would swirl around and create a lot of destruction in their paths.

A savvy innovator trying to move through an opportunity tornado uses speed, relationships, and aggressiveness for success. Some innovators love this environment because they are free to generate and execute on ideas. Others find it frustrating by the over-politicized environment. Don't expect that simply following the company's processes will work. Instead, find executives with the power to help you navigate the informal systems to get funding and smooth the path to execution.

The Opportunity Engine

This is the final environment and place most companies desire. They are high on both the idea generation dimension and the opportunity processing dimension. They might have numerous process experts running around, but not necessarily. Engines strike a balance between calculating risks and managing long-term emerging opportunities with near-term operations challenges. They have a range of ideas brewing at different stages lead by trained, savvy innovators who are usually leading or working in small teams. Opportunity engines regularly review progress for the best ideas, determine how to best move the opportunity forward, and are not afraid to kill ideas based on solid market and customer data. Opportunity engines are generally a great environment for a savvy innovator since they have a rule book that provides guided flexibility.

Amazon, 3M, and Procter & Gamble are good examples, but you don't need to be a Fortune 100 company to be an opportunity engine. Take for instance, a privately owned, billion dollar medical-device company in Portland, OR. I once asked the product manager how his company approaches innovation. How does it gain customer insight, evaluate new opportunities, and make decisions? From each response, it was clear the company had a solid vision, fostered new thinking, and knew how to transform concepts into real business opportunities—all with a clear customer and market focus. The product manager didn't use a single buzzword from Lean, Six Sigma, staged development or any other process. He didn't need to. The company had become an opportunity engine through enlightened business sense and simple innovation processes.

Navigating Ideas Through Turbulent Innovation Waters

These environments are not limited to large companies. They exist in the smallest companies as well as startups. Sometimes there are several environments in one company. I remember a large tire company with an advanced R&D group that was a tornado, but before they could commercialize an idea, they had to place it in one of the business units responsible for manufacturing and distribution. Each business unit was a siphon that would only accept ideas already proven in the market. This created frustration all around.

And consider Google. What is its opportunity environment? Most would say an opportunity engine, but I believe it's a tornado. It has numerous innovation projects, but how many will drive the success of the company? Only time will tell, but I speculate that if Google's core advertising products start to decline, it will be forced to become an engine and implement more rigorous idea screening for its stockholders. It's OK to be a tornado when you have billions of dollars in the bank, but most companies don't have this luxury.

I knew an innovator at one media-technology company who was a master at getting funds for his ideas as long as it was a tornado. Once the company started putting in more processes to prioritize opportunities and use a more systematic approach, he struggled with adapting and left after one year. Yet others began thriving because now they had the structure of a siphon that they needed. Savvy innovators know their environment and how to adjust accordingly to navigate an idea through that environment. For instance, working in a clog environment requires a very different strategy than working in a tornado or engine environment.

SAVVY IDEA: CREATE YOUR OWN IDEA ENVIRONMENT

If you are in an environment that does not embrace ideas or has processes to evaluate them, why not create your own! A simple way to get started is to form a small team and work through the 30-Day Action Plan in Part III of this book. With the right methods and activities, you can show others in your company how to approach innovation. It's not necessary to announce what you're doing. Your work and results will talk for you. Once you have success, then share your approach and show others how to do it...one idea at a time. Company cultures can be complicated. The best way to influence is usually by example.

The next chapter will focus on the executives who hold the innovation resources and will ultimately determine the fate of your idea and possibly your career as an innovator.

CHAPTER 3 REVIEW—CORPORATE OPPORTUNITY ENVIRONMENTS

Corporate politics and environments can be complicated, but if you're able to clearly understand and accept the culture, you can put your time and energy into the right activities and get your ideas heard. As your influence and reputation grows, you can start to affect deeper and broader changes to build a more innovation-friendly culture.

Five Savvy Success Strategies

As you hone your skills, practice these five strategies to understand and navigate your corporate environment.

1. **Don't get frustrated by the politics.** If you're not part of the political game in your company it's easy to get irritated when you see others trying to influence decisions or selling their own agendas. But these are skills you must also build. Rather than get frustrated, get in the game and learn how to play.

2. **Flow with your company's current.** No company is perfect. I always hear complaints that executives don't listen to ideas or that the company emphasizes revenue instead of innovation. This is the reality. Trying to change the entire culture is like a fish trying to swim upstream. I'd be great to correct all of the issues thwarting innovation, but you probably don't have the authority, resources, or time to fix systemic problems. Instead, to compensate for the challenges, focus on learning what type of environment does exist and adjust your style to the corporation since it will not adjust its environment to you.

3. **Understand the innovation portfolio and budget.** Without a budget, nothing gets done. If your company does not have funds set aside specifically for new opportunities, then the funds must come from current operations budgets, which means taken from another project. Knowing where the fund-

ing comes from helps you focus all elements of your opportunity proposal.

4. **Know your company's strategy and objectives.** Any idea that requires significant resources must be aligned with the company's vision and goals. Sometimes this information is easily available, but it can be obscured and must be learned by working directly with executives. Do your homework to determine what the real objectives for innovation are and how your idea fits with the corporate goals.

5. **Know your company's history.** Failed attempts at innovation leave a powerful imprint on the corporate memory. The CEO of a wood products company once told me that his company couldn't find ideas to help it grow. Two years earlier the company had been excited about a new type of door it had launched. The company spent a lot of money, but ultimately the project failed. After the failed door project, the company didn't even consider ideas without an obvious path to revenue and profit. They'd tried to innovate but got bitten. Review the historical initiatives to understand the whole picture. Find out how they originated and what led to their failure or success.

BOB AND NANCY (CONTINUED)
MedCo raises its corporate defenses.

Both Bob and Nancy are getting frustrated. She is tired of asking questions that he can't answer. He is getting tired of answering questions that don't seem relevant. He thinks they should focus on the product and not a bunch of numbers. Nancy just said, "I think your idea has potential, but let me ask you a couple of more questions."

Bob: *OK. What else can I answer about the Wireless HM-3000? That's just a name my team came up with. I think it fits though.*

Nancy: (Hmm. I'm sure marketing won't be happy with that.) *So who else have you shared this idea with? Have you run this by the sales team?*

Luckily, Bob did have a quick discussion with Ted, MedCo's head of sales, but he thought that doctors were happy with the current product and wouldn't want to change. Ted told Bob, "Wireless is too complicated. What if it fails?" Bob listened as Ted talked about all the times he unsuccessfully tried to get on a Wi-Fi network at airports. Bob figured Ted didn't get it.

Bob: (What should I tell her? If I say Ted doesn't support it, that won't look good for either of us.) *I've run it by our CTO. He loved it and doesn't see a big problem with development. I also ran it by sales and they are thinking about it.*

Nancy: (Of course the CTO loved it...it's new technology. I've heard enough. I need Bob to go away and come back with better answers.) *OK. Let me tell you more about what I need to see. For new products, I need to see revenue by the end of year two and cash flow positive by the third year of investment. I'd also like to see margins of at least 50%. I want to be clear on who the target market is and see data that can support your revenue forecasts. Can you send me a proposal that provides this information?*

Bob: (Wow! This is innovation. How can we know all of this now? But she is the CEO.) *No problem. I'll have that for you next week.*

Nancy: (Sounds too fast, but I'll see if he can do it.) *Great. Let's put something on my calendar. Sounds like you could use some marketing data. Check with Tanya in our market research group. She has access to a lot of industry reports. I look forward to hearing more.*

Bob: (I can't wait to get started on development.) *OK. Thanks for the time. See you next week.*

WHAT'S GOING ON?

MedCo has a typical Opportunity Siphon environment. Nancy needs to see a lot of data before she commits to funding Bob's idea, but he still hasn't been able to provide the real answers she needs. She's trying to be supportive and thinks the idea has merit, so she allows the HM-3000 to pass the Five-Point Inspection and gives Bob the go-ahead to move forward with developing an opportunity proposal. She's even pointed Bob to resources for help. She could have just as easily said she'd think about it and let it die a slow death.

Bob, being new to all of this, has only given himself a week to develop an opportunity proposal that will answer Nancy's questions.

Will Bob be able to get Nancy's approval and the resources he needs to transform his idea into a real business opportunity?

(To be continued...)

Chapter 4

THE CORPORATE EXECUTIVE MINDSET

Various Personalities And Their Needs

Corporations don't buy ideas,
people do.

Rate Your Current Skills

- How well do you know the people responsible for making and influencing the decision to fund or kill your ideas?

- What are the primary objectives that each corporate decision-maker must achieve to look good to his or her boss?

- What are the individual attitudes about risk and the level of analysis and confirming information each executive needs before making a decision?

- What are the recent projects that each decision-maker supported, or didn't support, and why?

- How well do you know what is personally at stake for decision-makers if they say "yes" to a funding request?

Introduction

While executives and innovators often work in the same building, they usually live in different worlds. Innovators typically live in the conceptual world of ideas, product development, features, technology, prototypes, and creation. Executives typically live in the analytical world of market trends, hurdle rates, profit margins, revenue, portfolios, and risk management. For example, usually an engineering leader and a CFO think very differently about innovation. They might not even agree on a common definition of the term.

No matter how much a company standardizes its innovation process to bring these worlds together, selling ideas to corporate decision-makers comes down to human behavior. When an innovator sits across from an executive to ask for resources, success is based as much on his ability to communicate, negotiate, and persuade as it is on any hot technology or cool product features. Consequently, innovators must fully understand the mind of corporate decision-makers and learn how to effectively sell their ideas to those executives who control the resources and their destiny.

Corporate Executives Come In Two Flavors

When I use the term *customer* most people think about the potential buying customer for a product or service. But a savvy innovator realizes the first customers are corporate decision-makers—the executives who say yes or no for approving precious resources.

Generally they fall into two basic categories:

Funding executives, who have the authority and budget to say yes to fund your idea. Often this is the CEO. In larger companies, with more distributed budgets and authority, this may be a general manager of a business unit or a special position set up just for funding early-stage opportunities, such as a chief innovation officer (CIO).

Advising executives, who don't have direct authority to approve funding but will greatly influence the decision. They might be functional vice presidents in sales, R&D and marketing and, in larger companies, they may be the chief technical officer (CTO) or a chief marketing officer.

Funding executives will seek confirmation and advice from these other decision-makers to determine whether your idea merits corporate time, attention, and resources. In almost cases, one of the primary advisers to the funding executive is a trusted financial adviser. This may be the CFO or other senior finance person, depending on your company size and structure. These advising executives are important to the innovator because even though they can't say yes, they can say no and then make it difficult for the funding executive to say yes.

While you ultimately need the funding executive to say yes, you'll always need the support of anyone that can say no. Throughout the book, I'll refer to any executive who can say yes or no to your idea as a decision-maker, funding executive, or just plain executive. Let's take a deeper look at the types of personalities you'll encounter.

The Corporate Decision-Making Persona

While every executive is unique, they typically share characteristics. They each worry about staffing issues and achieving results and hitting financial goals. But they also have additional characteristics that contribute to common stereotypes, such as:

Having short attention spans. The higher executives are in a company, the broader their responsibilities become and the less attention to detail they can have. This leads to short attention spans and the need for short, focused communication.

Being broad but not deep. To stay informed, executives must be lifetime learners. They read everything on every topic. They know a little about everything, but are not necessarily deep on any topic unless it's an area of passion or directly related to the business.

Taking pride in their success. They've worked hard to achieve their large paycheck and status in the company. They're not going to let an ill-conceived concept interfere with their lives.

They worry about their bosses too. Executives know that their success is based on making their boss successful. Even CEOs must make their board of directors happy because their future depends on it.

They can seem aloof. With all of the employees, customers, investors and vendors trying to get their attention, executives can seem distant. After all, everyone knows them, but they may know very little about the hundreds or even thousands of people who report to them through many levels.

Top executives must also be able to make tough decisions. No one wants to be seen as a "waffler." The 2004 presidential election was a good example of the devastating effects that changing an important decision can have on an executive's career. John Kerry reversed his decision to support the Iraq war, and George Bush pounced on the perceived character flaw to position Kerry as someone who couldn't make hard decisions.

Sometimes these characteristics are so prevalent it's difficult to get past stereotypical personas and not think about executives as corporate robots instead of as human beings—with family issues, financial concerns, and personal insecurities like the rest of us. Stereotypes are further intensified with the most senior executives because they achieve an almost celebrity-like status, where everyone in the company comes to know them by their carefully crafted public persona and not the real person behind the scenes.

These Characteristics Can Create Problems For Innovators

Each of these executive characteristics can conflict with the personality traits and needs of innovators. Thus, communication problems can arise when discussing ideas. An innovator sharing his deep knowledge might frustrate an executive trying to understand with his broad, but shallow knowledge base—and vice versa. An executive with a short attention span might frustrate an innovator trying to have an extended conversation about various options. Contrarily, making a fast decision without supporting data is extremely difficult for an executive whose whole reputation hinges on an ability to make good decisions without waffling.

Four Corporate Executive Mindsets

While these characteristics provide insight into behaviors, executives often behave quite differently when asked to fund a new idea. Saying yes or no is based largely on the executive's attitude toward new ideas. Some say no to nearly everything and others say yes quite easily, based on just a short conversation. These behaviors may be because of past experiences, part of the executive's DNA, or frankly just how close he or she is to retirement.

Regardless of the root cause, a savvy innovator knows how essential it is to learn the decision-making behavior of any individual executive. Having this insight informs your approach when seeking support to fund your idea. This saves you time and effort and dramatically increases your odds for success.

Figure 4.1 shows four categories of executive personas based on two dimensions: 1) Open to Ideas, and 2) Data Needs.

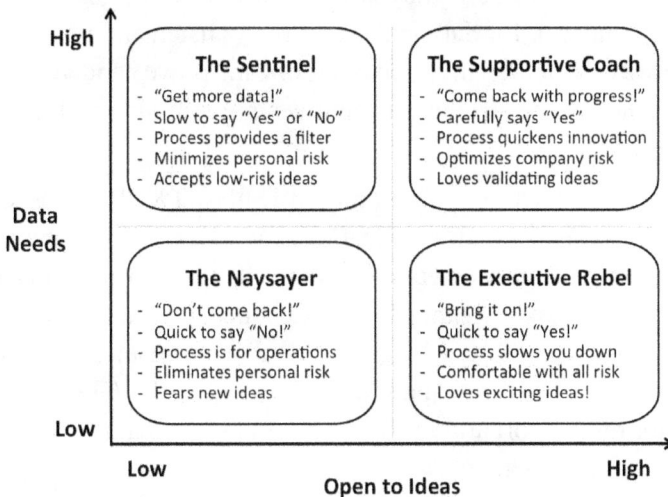

The Sentinel
- "Get more data!"
- Slow to say "Yes" or "No"
- Process provides a filter
- Minimizes personal risk
- Accepts low-risk ideas

The Supportive Coach
- "Come back with progress!"
- Carefully says "Yes"
- Process quickens innovation
- Optimizes company risk
- Loves validating ideas

The Naysayer
- "Don't come back!"
- Quick to say "No!"
- Process is for operations
- Eliminates personal risk
- Fears new ideas

The Executive Rebel
- "Bring it on!"
- Quick to say "Yes!"
- Process slows you down
- Comfortable with all risk
- Loves exciting ideas!

Figure 4.1: Four types of executive innovation personas

These two dimensions range along a spectrum from low to high. For instance, executives will range from not being open to new ideas to being very open. For those who aren't open, you'll need to build a network of support and plant seeds so they'll start to believe the idea was theirs in the first place. Similarly, executives will range from not needing any data to needing a lot before making a decision. To gain the support of executives who need mass amounts of data, you must develop an opportunity proposal with enough relevant data that will calm even the most risk-averse executive. Depending on how a decision-maker leans in these two dimensions and factoring in the following four personas, you'll be able to anticipate and counter the reasons why they might say yes or no far in advance of the first funding meeting.

The Naysayer

These executives usually start and end by saying no to almost every idea. Naysayers are not open to ideas and pay little attention to data because they typically believe such data is inaccurate. They draw quick conclusions from previous experience, which may include failed innovations. Naysayers are protecting their own status and believe most ideas fail, so if they say no, they'll be right most of the time. They may ask for data, but won't really listen and hope you'll just go away. Whether they have significant power and resources or few at their disposal, these executives are still able to influence investment decisions.

I remember one COO of a company in Dallas, TX. The new CEO was trying to identify new applications for industrial solvents. The development team presented more than 30 ideas from consumer drain cleaning to toxic cleanup. After each idea, regardless of the enthusiasm, the COO retorted, "We can't do that because…" Or, "XYZ competitor would kill our margins." Eventually, the CEO stopped looking for his support and approved several ideas that moved to the next stage of development.

Working with Naysayers is difficult, but ultimately you'll need their support, so stay close to them. You'll rarely get a yes, but try keeping them from saying no. If the Naysayer is the CEO, do the best you can to present your case in the most data-rich, compelling way possible and build a groundswell of support with the CEO's most respected team members.

The Sentinel

These executives' first reaction is usually "prove it to me," but will default to no if proof isn't readily available. Sentinels don't easily embrace new ideas. They need lots of data before making a funding decision. They're protecting the status quo and company assets. They want ideas to be well-researched with risk removed before considering the idea. Sentinels can be quick to point out the flaws in your thinking and the gaps in your data. They'll require you to go back for more data until it's clear the idea has potential or you give up and go away. They do fine in slow-moving markets or where there are few competitive forces. Sentinels have difficulties when markets change quickly or an aggressive competitor appears.

I've worked with many Sentinels in my career. The last one was the general manager of a cloud-based software company. They had dozens of R&D projects going and not a lot to show for them. Since the GM came from the consumer-goods market, she was extremely market research oriented. Before approving any more funding, she required her team to prove with statistical accuracy that any specific application would be successful. Out of 23 projects, only two made the cut. In this case, she had to make some tough decisions to make the company stable.

Working with Sentinels creates a different challenge. With a Naysayer you know exactly where you stand. With Sentinels you may feel like you're making progress, but they always want more—more data, more proof, more time. A savvy innovator will have persistence and gain support from the finance team and other advocates who can vouch for the plan.

The Executive Rebel

These executives love innovation and aren't afraid to show it. They're open to ideas and rank low on the need for data. Sometimes they have the formal power and resources to fund ideas and sometimes they don't. Executive Rebels may be innovators themselves with a history of shaking up their past companies and departments. They're often involved with various ideas of their own while helping others. If they don't like an idea, they're quick to dismiss it. Executive Rebels often form cliques with other innovators to identify and execute new opportunities. They are comfortable with risk and often shun formal processes. In their minds, processes slow everyone down. Rebels rely on their gut, direct customer interaction, and personal interest more so than data. Some environments need Executive Rebels to shake things up, but they can also wreak havoc in unprepared organizations.

I worked with Sarah, a senior VP of product management for a test equipment company. She had a business-development background and loved a good idea. She'd work directly with the company's top customers and take new ideas directly to engineering, where she'd convince them to develop prototypes. She drove sales crazy since she'd promise things they maybe couldn't deliver. And the engineering team was never sure what was high priority. Some products became successful. Others just sucked up R&D time. The results? Mixed. She left the company after three years when it became more difficult to get projects through.

Working with Executive Rebels can be a wild ride. If your ideas align with their interests and passions, getting their support is easy. However, if your idea is not of interest or if they don't like you for some reason, they can be difficult to work with. A savvy innovator appeals to their passion (often their ego) to get them excited about an idea.

The Supportive Coach

Supportive Coaches balance being open to new ideas with the desire to see lots of data. These executives love a good idea and want to see a range of opportunities being developed simultaneously to balance risk for the company. They actively listen to new ideas, but also ask tough questions. Supportive Coaches use their power to help innovators succeed, but expect due diligence from their innovators before saying yes or no. "Come back with progress!" is their mantra. These executives are often found in more mature innovation environments and may be the main drivers of innovation processes.

Frank was a Supportive Coach in Los Angeles for a company that had identified an opportunity to develop a direct-to-consumer offering for its audio speakers. He was hired to start and staff a new division and make sure the new sales channel was successful in the market. He needed many new ideas for products, marketing programs, and operations improvements. He set goals for each area and established some simple filters to prioritize ideas. When innovators had ideas, he provided resources to execute them. If they struggled, he'd coach them through the problem or find experts to help. Frank accomplished his mission with this style and went on to lead all of the company's innovation efforts.

Supportive Coaches are the easiest for most innovators to work with since they understand that ideas are merely starting points. They provide the time, resources and protective cover that innovators need to work on their ideas. Coaches are not pushovers though. They'll still err on the conservative side where markets don't yet exist and will require consistent progress and urgency to mitigate both technical and commercial risk.

Success Equals Meeting Personal Needs

Whatever decision-maker you encounter, always remember that executives care mainly about what you can do for them, both personally and professionally. A savvy innovator understands a decision-makers' needs and meets them as effectively as possible.

SAVVY IDEA: ASSUME EVERY EXECUTIVE IS A 4TH-GRADE GENIUS

Understandably, it's crucial to be able to quickly explain an idea and its benefit to decision-makers. Frequently innovators walk away from a presentation thinking that executives "just don't get it." Often the innovator is right. This happens when the innovator goes too deep into the detailed aspects of the concept that lose his audience. Sometimes the innovator assumes knowledge the audience doesn't really have.

To remedy these issues, bridge the gap between their current knowledge and the information you're sharing. Think of your executives as 4th-grade geniuses. 4th-grade geniuses understand fast, learn fast and ask questions. However, they haven't been thinking about your concept as deeply as you have, so they don't know what you know. To explain your concept quickly, for example, try telling a story of how a customer might use your product instead of an existing product. Complement this story with a picture if possible. Instead of spoon feeding executives a long presentation, make it short and focused, and be prepared with time to answer key questions.

In the next chapter, we'll explore communication challenges and the languages barriers between innovators and different executives.

CHAPTER 4 REVIEW—THE CORPORATE EXECUTIVE MINDSET

When I say that corporations don't buy ideas, people do, I mean it's important to remember you're working with different types of decision-makers and their personalities. You must know how to effectively sell your idea to each type. Savvy innovators take time to understand executives from a human perspective and see them as partners and allies. It's a symbiotic relationship that holds many positives. A savvy innovator also takes time to gather the right data and craft a targeted message. This helps build a reputation as a trusted professional, who's not just pitching ideas, but sharing information and opportunities.

Five Savvy Success Strategies

As you hone your skills, practice these five strategies to work successfully with corporate executives.

1. **Understand each executive's role and goals.** Titles like CTO or SVP of Product Management seem straightforward, but their specific role and goals are not always clear. For example, is the CTO tasked with identifying technologies to improve operations? Work with technology partners? Or something else? Is the SVP of marketing focused on branding? Channel development? Or new products? This will be critical information for targeting the specific message to sell your idea.

2. **Research your executives' backgrounds.** Innovators often spend more time researching technology for an idea than learning about the executives who control the future of their idea. Research their educational and business background and where they have succeeded or failed. For example, you may find that an executive launched an industry-changing product during his career. This information personalizes the executive and helps you find common ground to work towards the success of your project.

3. **Understand the corporate hierarchy and authority of the decision-makers.** Just as sales people know to ask if a customer has authority to make a purchase before making a sales pitch, savvy innovators realize each level of executive has a different level of spending authority. If you're asking for $1 million in funding, you will likely need to go all the way to the CEO, but if you ask for $25,000 to get your idea to the next stage, then you may be able to get approval from a vice president or director. Determine the highest level of yes you need, then work backward from there to know which executives to approach.

4. **Respect each executive and their needs.** Most executives are specialists in their own field, but may know little in others. If they don't immediately understand your idea, that usually means they don't have the right background. They may even ask seemingly naïve questions. You can build strong allies by helping them come up to speed on new technologies, markets or products.

5. **Approach the process as a partnership.** Forget the 'us-versus-them' mentality. You are the subject matter expert with a great opportunity. They are responsible for creating profit from new sources. You need each other! Executives are the gatekeepers to funding, but the more you treat them as partners with mutual goals, the easier it is to hold discussions and come to conclusions that meet both of your needs.

BOB AND NANCY (CONTINUED)
Nancy shows her executive personality.

Bob agreed to take a week to build an opportunity proposal for his Wireless HM-3000 heart monitor. He took Nancy's advice and reviewed an industry report that Tanya from marketing gave him. He worked out some numbers and developed a spreadsheet that he believed Nancy wanted to see on the financial potential for the product. He also finished the product requirements and developed a more detailed schedule with engineering estimates.

It's Monday morning, one week later. Bob is back in Nancy's office with a 53-page presentation.

Bob: *Good morning Nancy. I have some of the answers you were looking for from last week.*

Nancy: *Great. What do you have for me?*

Bob: (Turning to page 35.) *Let me first show you the new schedule after I worked out some of the details with the development team. I think we can beat the schedule I shared earlier and get this product out in 12 months at only $3.2 million.*

Nancy: (Great. Another Gantt chart.) *Let's talk about what you learned about the market.*

Bob: (Feeling a little deflated.) *OK. Here are the numbers I developed. Assuming we launch this product in 12 months, it looks like we'll have no problem hitting revenue of $35 million in year two going to $120 million by year four. Based on the costs I estimated, this will give us profit of $15 million in year two growing to $60 million in year four. This hits your 50% margin needs, and considering the $3.2 million for development, this should have a very high ROI.*

Nancy: (Clearly he hasn't had the finance team review his numbers. He's just telling me what he thinks I want to hear.) *OK. These are some big numbers. Where did you get these revenue forecasts?*

Bob: (Shoot, I was hoping she wouldn't ask.) *From the industry report that Tanya gave me. Wireless heart monitoring is expected to be a $500 million product category by year four. I assumed we would get about 20% of this market. I have no idea what the price would be yet. I asked the sales team for how many they could sell, but I haven't heard anything yet.*

Nancy: (Sigh.) *OK. Tell me about what you've learned about the target market you researched.*

Bob: (Turning to page 48, which shows a long list of contacts. I'm glad I spoke to the product manager about this.) *Well, I learned that the real customer isn't the actual patient. The customer is really the hospital that buys our monitors. Here is a list of our top 25 hospital customers. I think all of them will go for it.*

Nancy: (This is a start, but still doesn't tell me what I need to know.) *Ok Bob. Thanks for putting this together. I need to get on a phone call. Leave this with me and I'll take a look at it.*

Bob: (Hhmm…Is she blowing me off? Or does she really need to think about it?) *Um. OK. When is a good time to follow up with you?*

Nancy: *Let me get back to you.*

Bob slinks out of Nancy's office, completely deflated. He had high hopes for getting her approval to move forward, but now does not where he stands.

Nancy thinks Bob's idea has potential, but needs someone she can trust to analyze the opportunity and provide the data she's looking for. She sends a note to Bob the next day.

"Hi Bob, thanks for meeting yesterday. I appreciate your work on the proposal. I'm going to ask Keith in strategic planning to take the lead on analyzing the idea you proposed. Please send him the information you have on the project. Best regards, Nancy"

Bob is stunned. He feels betrayed even though he knows not to take it personally. He wonders how he was supposed to find all of the info she wanted. Her questions were all about finance and not about the product at all! She wants information that doesn't exist. I'm not a market research specialist. Isn't that Tanya's job?

WHAT'S GOING ON?

Nancy is still not happy with Bob's proposal and as much as she was striving to be a Supportive Coach, her Sentinel executive personality began to show. She was disappointed that Bob had still not developed the opportunity proposal that she needed, to the point she's ready to move the project to someone else.

This lack of communication between them has started to have serious repercussions on their ability to find common ground on Bob's idea. It's also started to jeopardize Bob's standing with Nancy.

Can Bob recover from this latest development and salvage his idea—and his reputation?

(To be continued…)

PART I CONCLUSION

Corporate innovation comes with many challenges along with its rewards. The goal of Part I was to examine some of the situations you'll confront. You'll need to accurately assess your own skills as an innovator, understand the type of idea you're working on, be aware of your environment, and get into the mindset of the executives you'll encounter along the way. A savvy innovator keeps these factors in mind when formulating the plan to move an idea forward.

Next on the path for gaining support and funding for your BIG idea is learning how to evaluate and communicate the value of that idea. We'll cover these topics in Part II of *The Savvy Corporate Innovator*, titled, A Corporate Innovation Translation.

PART II

A CORPORATE INNOVATION TRANSLATION

A Guide To Key Terms, Concepts, Questions, And Answers When Facing The Executive Inquisition

INTRODUCTION TO PART II

Ideas can't speak for themselves, so you must give them a powerful voice.

At some point you'll be face-to-face with decision-makers to defend your idea and ask for resources. These discussions can be a simple hallway conversation or a formal funding meeting with everyone in the room—*the executive inquisition.*

To achieve a favorable outcome in each case, a savvy innovator must clearly address the key issues that always arise when discussing new opportunities.

In Part I of *The Savvy Corporate Innovator*, we explored the world of innovation—what it entails and how to navigate it. In Part II, we'll address the five most common challenges you'll encounter while attempting to get support for your BIG idea.

Chapter 5

THE BUSINESS LANGUAGE OF INNOVATION

Defining Fuzzy Terms And Phrases

Selling an idea requires good communication. Unfortunately, there are many different languages in every company.

Rate Your Current Skills

- How well do you know the key terms used by each decision-making executive?

- What are the exact definitions of risk, value, customer and market as applied to innovation?

- How well do you adapt your vocabulary depending on the person you're talking with?

- How comfortable are you translating a term that someone is using differently than you expect?

- What do you do when a conversation is straying from a mutual understanding to get it back on track?

Introduction

Remember, your goal for the next 30 days is to develop a compelling opportunity proposal that results in first-stage funding. But first you must face and pass the executive inquisition. This is typically a formal review meeting where executives from each function of a company will ask pointed questions about your idea. You must explain your idea and effectively answer their difficult questions about why your idea is worth pursuing. Since new opportunities affect everyone in the organization, innovators must be ready to talk intelligently with executives from every field including marketing, finance, operations, engineering, and sales.

However, every function in a company has its own language. Marketers speak in terms of brands and positioning. Software developers speak of releases and bugs. Sales people speak of quotas and closing. And CEOs speak of core competencies and shareholder value. You're probably fluent in your own field's language. But you'll have problems if you only use that language in a funding conversation. It's essential to learn the other professional languages or you risk being labeled as an outsider, quickly lose their respect and increase the odds of hearing "no." A savvy innovator becomes fluent in every field's language especially as it relates to funding your idea.

The Basics Of Communicating Successfully About Innovation

When speaking with executives, their world and unique language can seem foreign if your background and discipline are different than theirs. While it's often a technical person who feels foreign in the world of sales, a sales person with an innovative idea can feel equally lost in the world of technology. For example, to most people the word 'sprint' means "a short, fast run." But to software professionals, a 'sprint' is a "short software development cycle that results in demonstrable features."

How we interpret words during a funding meeting determines whether we reach a mutual understanding of a new opportunity or end up frustrated and going nowhere. A savvy innovator knows this and adapts to the role, field of business, and background of the other person. Think about how you communicate with decision-makers. Is the conversation in your language or theirs?

"When I said you should improve security
I meant 'put a latch on the door'."

Are We Really Communicating?

The French have a superbly descriptive term that has no exact translation in English: *quiproquo*, pronounced kwee-pro-kwo. This shouldn't be confused with 'quid pro quo', which describes an exchange for goods. No, 'quiproquo' is when two people have a full and often lengthy discussion using the same words, but in a sense having two entirely different conversations—in essence, a misunderstanding.

A QUIPROQUO EXAMPLE

While at a tradeshow in New Orleans, some colleagues and I met with a potential distributor that wanted to carry Motorola's products in their catalog. "Great," we said, "Send us a business plan and we'll take a look." We met again the following week and they presented their "business plan." It had lots of pictures showing how great they were, a history of their company that included a video from the president,

---

and one page of revenue projections for our product line. My GM looked at me, rolled his eyes and left the room.

We'd had a quiproquo moment about a completely different definition of "business plan." We expected a plan showing their marketing activities for the product line, training for their sales people, and how they'd manage returns and customer service. They assumed something else. The result was they didn't get Motorola's business.

The quiproquo problem is universal. For instance, in China I was working with a large developer of electronic components. The head of their innovation team asked me to critique 20 funding presentations for their American corporate executives. What I found amazing wasn't a translation problem in their English; rather, it was in the business language of innovation. Just as I often see with English speaking innovators, the Chinese presenters were unclear on how to present concepts like markets, customers, and finances in terms that executives needed to hear.

A Savvy Innovator Never Makes Assumptions

An R&D expert in the field of bio-engineering wouldn't expect a marketing executive to know the intricacies of the latest application of genetic markers or engineering issues. As an expert in any field, it's easy to forget that other, non-experts, don't understand many of your terms or concepts. Buzzwords and acronyms are fine with peers and colleagues, but when communicating with people from other fields, it's important to use language they will understand.

One field where I often forget this basic tenant of communication is in the field of innovation. Words like *disruptive, radical, open, ideation, concept, experiment, hypothesis,* and *exploratory* can confuse people (like prospective clients) who don't know our world. A glazed look or two usually jolts me back to remembering that others may not be clear on these terms and especially their context. Anytime you use a term that has a unique application to your work, make sure you recognize and clear up any misconceptions.

The Savvy Innovator Dictionary

Below are seven fuzzy terms often used when discussing innovation. But they mean different things to different people. To reach a mutual understanding, you must discover what the word means to the other person. Consider 'customer' for example. A simple word used constantly. But unless you know exactly what the other person means when she says, "Tell me about the customer for your idea," you'll likely use your own definition and experience a quiproquo moment.

Innovation. A term with many shades of meaning. Designers may define it as 'creativity' while technical people may define it as 'new technology' or 'inventions'. The term innovation may evolve to a more consistent definition similar to how 'quality' became more precise as processes were formalized. But for now, make sure other people understand what you mean exactly.

Savvy Innovator Definition: To systematically identify and execute opportunities that enhance and develop new products, attack new markets and define new approaches. Innovation starts with ideas. It ultimately leads to improved business results.

Idea. Executives often say, "We need more ideas around here!" Because the term is so broad, it's essential to find out exactly what people mean. Do they mean big ideas that will disrupt the industry? Ideas to add new features to existing products? Ideas to improve operation? Ideas to find new customers? Or all of the above?

Savvy Innovator Definition: Any captured thought that may create an opportunity for a new product, market, or approach to doing business for increasing revenue, profit, or competiveness. Ideas lead to innovation. A 30-Day Idea is one that passes the Five-Point Idea Inspection discussed in Chapter 2.

Customer. Depending on their role and situation, people use this term differently. There are 'current customers' and 'potential customers'. Sales people refer to someone who is a direct buyer, such as a purchasing agent or distributor. Technical people refer to a product user. People also refer to companies as 'customers', but since companies don't buy products (people do) it's important to clarify who inside the company is the actual customer.

Savvy Innovator Definition: The person who actually purchases or directly participates in the decision to purchase a product or service. In cases where money doesn't change hands, such as a non-profit or an internal application, then the primary decision maker is your customer. For your ideas, corporate executives are your customers.

Data. It's often assumed that data means statistical information from either an industry report, quantitative market research or other authoritative source. But, data is not just numbers—it comes in many forms. For example, if you interview customers and document the results, that is data. Summaries of competition, a reaction from a key supplier, or an expert's opinion are also important types of data. Data also varies according to who requests it. Remember, a marketing executive is not looking for the same data that an R&D executive requires.

Savvy Innovator Definition: Any piece of relevant information used to clarify or validate your opportunity designed to gain support from funding executives.

Value. One of the most overused terms in business. So much so, it has become almost meaningless. Statements like, "We create more value for our customers!" are rampant. Like art, it's in the eye of the beholder and changes meaning based on each person's situation, role, and background. To some, it's 'customer benefits'. To others, 'new technology' or 'profit'. For innovative ideas, 'value' serves to fulfill customer needs.

Savvy Innovator Definition: Value refers to how much worth the customer places on your innovation compared to purchasing alternatives or doing nothing.

Risk. Simply put, uncertainty about the future. Executives are always concerned about different kinds of risk and the factors that create the risk. They see uncertainty in R&D schedules, customer adoption rates of new products, and market conditions. Problems often occur when you seek funding and their specific concerns aren't addressed to their satisfaction. Most innovators address risk too broadly with statements like, "We have the risk of a competitor coming out with a similar or better product."

Savvy Innovator Definition: The internal (company-based) and external (market-based) factors that create real or perceived uncertainty for success when pursuing a new opportunity. Each executive will have very specific and unique fears when considering risk.

Market. Even professional marketers can be fuzzy on this term. Some think of a product category like the 'refrigerator market.' Others think in terms of a technology or geographic region such as the 'x-ray market' or the 'European market'. But ideally, markets should be defined in terms of customer groups with common needs. For example, rather than defining a market by a product category (the refrigerator market), define it by what the customer needs; 'the set of consumers that need to keep food preserved.'

Savvy Innovator Definition: A set of customers who have common characteristics and needs. A target market is the set of customers to whom you plan to promote and sell a product or service.

Those listed are the most common, but there are many others such as: *financials, forecast, research, marketing, strategy, requirement, competition,* and *technology.* Bottom line, terms are used differently by different people and it's important not only to think about the terms you're using but also whom you're using them with. While I might believe markets should be expressed as 'customer needs', if the executives I'm speaking with believe they're in the 'refrigerator market', then I'll also use that term.

JOSE AND MARILYN: A DRAMATIZED COMMUNICATION CHALLENGE

The following story highlights just how common it is for innovators and executives to miscommunicate in the world of innovation. Read how Jose responds to Marilyn, his CEO, before learning how to speak the business language of innovation.

Jose was walking out of his development lab just as Marilyn was walking in. He had finished testing his latest version of facial recognition software that would help consumers log into any device, website, or application without needing to remember a password. Jose and Marilyn both despised passwords and couldn't wait to rid the world of a painful and flawed security mechanism.

"Hi Jose, just the man I was looking for. How's the project going? It's such a great idea that I promised the board of directors I would share some data on the innovation at the meeting tomorrow. They'd love an update on the market, the customers, the value of the new service, and any risks. Sorry for the short notice, but can you send me an update that covers those areas? I need it by morning. Just one slide per topic is fine."

Jose smiles, "Sure! I'd love to put that together for you." Wow! A chance to show off to the board. Jose's wheels were turning as he raced back to his office. He develops seven slides with the following content:

Idea slide: A picture of himself with a thought bubble that says, "Passwords suck!"

Innovation slide: 100 lines of his best software code. (As he copied and pasted the code onto the slide, he thought, yeah... this'll show them how innovative this is.)

Market slide: A mosaic of 24 devices, applications, and websites with a message on top that reads, "Never Forget a Password Again!"

Customer slide: Another picture of himself.

Value slide: A bulleted list of the features of facial recognition—even in low lighting!

Data slide: Jose was confused what to include here. He thought, "Hmm, I think she wants to see market data." So he includes several bullets listing how many people use passwords (which is like everyone!) and how many websites needed passwords (which was quite a few).

Risk slide: This was easy. He listed the top ten bugs he was working through.

Jose looked at his presentation with pride as he hit the send button to Marilyn.

The next morning, Marilyn received the email and looked through the presentation. She couldn't help but smile to herself. "Ah...Jose," she thought, "what a riot." She hit delete and created a simple, one-page slide of what the board really wanted to know.

To prove to her board that fixing the password problem was worthy of their attention—and money—Marilyn needed concise information about the idea from a *market* and *customer* perspective. Not from Jose's perspective.

SAVVY IDEA: TAKE RESPONSIBILITY FOR ALL COMMUNICATION

When you're presenting an idea to people from different functions—say at the executive inquisition—the language problem is exacerbated because people interpret words differently. If the CFO objects with, "I don't believe you can achieve a reasonable margin with this product," you must respond in the CFO's language and also be aware that others in the room might need clarification. It's your job to be certain everyone clearly understands both the question and your answer. You must read everybody's reactions at all times. It's not just one-on-one communication. A savvy innovator takes responsibility for ensuring everyone in the room understands the discussion and doesn't form opinions based on their own interpretations. Otherwise, someone else might take control, and the end result may not be in favor of the idea.

In the next chapter, we'll look deeper into risk and how you can address executives' concerns so they'll say "yes!" to funding your idea.

CHAPTER 5 REVIEW—THE BUSINESS LANGUAGE OF INNOVATION

Every innovator selling a BIG idea ultimately faces the executive inquisition. A savvy innovator takes responsibility for the whole discussion and uses the business language of innovation that each executive needs to hear. Words are chosen carefully to avoid any potential quiproquo situations.

Five Savvy Success Strategies

As you hone your skills, practice these five strategies to successfully speak the business language of innovation.

1. **Learn the language from native speakers.** Whatever field you are in, spend time to get up to speed on the language of other functions by finding a tutor. Most people are delighted to share their expertise and get satisfaction teaching others. Just be respectful of their time and offer a mutual exchange. If you can't find a tutor, consider taking classes in subjects outside of your field.

2. **Consider any cultural differences.** Many environments will include people from different countries like China, India, or the UK. Be aware of word differences. For example, during one trip to China, I asked my colleague, Sheng Qun, to bring me back a hamburger for lunch. In China, all fast food sandwiches are known as 'hamburgers', so she brought me back a chicken sandwich.

3. **Seek a translator.** If you haven't mastered the business language of innovation, then use a translator. Maybe bring a financial analyst with you to a meeting with the CFO. If you're meeting the CMO, bring a marketing person with you, etc. Your goal is to have a clear conversation and mutual understanding, not to pretend you're an expert at every function.

4. **Bring outsiders into your circle.** When you have discussions with peers using special lingo and acronyms and people outside your world enter the conversation, they're usually uncomfortable if they can't follow the discussion. Look out for these situations and translate when needed so that the outsider feels welcome.

5. **Don't be dogmatic (please).** I love a good debate on the differences between similar terms like goals and objectives, but an argument never helps a situation. Even if you believe strongly in your definition over someone else's, stop the debate and focus on the outcome you're trying to achieve. If the definition is really that important, then work through it, otherwise save the debate for a fun lunch conversation.

BOB AND NANCY (CONTINUED)
Bob realizes he needs to learn a new language.

Bob felt dejected after Nancy's email saying Keith was going to take over analyzing the Wireless HM-3000. Bob couldn't stop thinking he'd lost control and that Keith would never give it the attention it deserved. Bob knew he could do better with a little coaching.

After some more thought, Bob realizes he may not have understood Nancy completely and wants another chance. Clearly he wasn't meeting her needs. Bob sends back an email in response to her note.

"Hi Nancy, I thought about our discussion last week and I'm happy to work with Keith on an update of the proposal. I know there was key information missing. I'd like the opportunity to take the lead on developing the update. I really believe in this idea and want to make sure you're comfortable with the risk. If I can get a little more guidance, I'm sure I can develop a proposal that includes the right information for you to make a decision. Thoughts? Bob"

Nancy reads Bob's email and appreciates that he wants to learn and grow and that he's passionate about the idea. She responds back.

"Bob, OK. I'll be traveling for the next four weeks, but set up a meeting for when I return. Take the next 30 days to really assess your idea. Hone in on answering the following questions:

- How can we prove that what you're proposing is something customers really want?

- What is the market potential for this product?

- What is a realistic revenue forecast for the next three years?

- What are the real risks involved and how can we remove them?

- What resources do you really need to get the project to the next milestone?

I need some innovative ideas to share with the board of directors. If the HM-3000 looks good, I may share it at the upcoming meeting. Nancy"

Bob is excited, nervous, and confused. He thinks, 'I can't screw this up, but her questions seem like she's asking me to regurgitate what I already told her. What does she mean by the market vs. customer? I already told her about the risk, what more can she possibly want? Maybe I do need some help here.'

Bob decides to call Keith, the market analyst.

Bob: *Hi Keith. This is Bob in engineering. I could really use your help.*

WHAT'S GOING ON?

Bob is an expert on technology, but has read enough business books to believe he was clear on all of the terms Nancy used: market, revenue forecasts, and risk. However, he has to be certain that he can answer her questions appropriately in the next 30 days. He's doing the right thing by reaching out to Keith to learn MedCo's specific language and what needs to be included in his proposal.

Can Bob learn fast enough to recover from his earlier attempts to sell his idea to Nancy?

(To be continued...)

Chapter 6

THE RISK CHALLENGE

Addressing The Risks Associated With New Opportunities

Every opportunity comes with many types of risk. The problem is in knowing which risks executives will accept.

Rate Your Current Skills

- How intelligently can you discuss the risks associated with your ideas?

- How would you describe the key factors that contribute to risk in terms that decision-makers want to hear?

- What separates executive risk factors from corporate risk factors?

- What are the risk factors that each executive worries about most?

- How do you make the risk of funding your idea acceptable to decision-makers?

Introduction

Investing in new opportunities presents an interesting paradox. If companies under-invest in new opportunities and focus solely on existing products and markets, they risk long-term failure. Polaroid, Schwinn, and Xerox are prime examples of this. On the flip side, if companies over-invest in new opportunities, they risk never seeing a pay off. Such as with Microsoft's heavy investment in Zune or Intel's investment in streaming video.

To achieve long term goals, executives must decide on the best opportunities and take the risk of being wrong. Often they have little more than fuzzy data to support their decisions. In the process they put their personal and corporate well-being on the line.

There is no way to eliminate risk when investing in innovative ideas. A savvy innovator knows the importance of strategically communicating and addressing the associated risks so that funding executives say "yes" and other influential executives don't say "no."

Risk Comes In Two Forms: Personal And Business

Innovation is simply 'managed risk'. And another word for risk is 'uncertainty'. Unfortunately, nothing about innovation comes with certainty. When an idea is conceived, all variables required to calculate any financial return are merely estimates (most might say "guess-timates")—from the overall size of a target market down to the cost of a product manual. As a savvy innovator, you know it's impossible to eliminate the business and personal risks related to your idea, so you focus on convincing executives to accept all the risks so they will approve the first stage of funding.

There are two major categories of risk in innovation: 1) Business Risk, and 2) Personal Risk.

Text books are ripe with information about the business risks involved when developing new products and entering new markets, but few books address the personal risk concerns that pervade every executive decision.

THE RISK CHALLENGE: THE 70% EXECUTIVE

Karl was an executive of a Japanese test equipment firm. He was looking for new growth opportunities, completely outside the current business, that met certain criteria: generate $30 million of revenue within three years, show profit margins better than 40%, etc. These were typical. But Karl wanted more. He told me, "Each opportunity must have a certainty of success at 70%." I thought to myself, "That's not possible in the early stages. How can we possibly assign that level? But knowing how important it was, I asked, "What exactly would give you a 70% level of confidence?" This lead to a discussion of the specific risk factors he was worried about, including the pressure he was under to be successful with the opportunities. Eventually, we agreed on a process the team could use to track progress and the milestones they'd achieve at each stage of development. By defining specific activities and having clear metrics for success, we agreed a confidence level of 70% could be achieved.

Not every executive will be so specific, but they will have their own criteria for deciding whether the risk is acceptable and will have specific steps you need to take to mitigate these risks. Your challenge is to determine those steps before you face the executive inquisition.

Personal Risk Depends On The Decision-Maker

To get support from executives, you need a strategy to address the risk factors that executives care about most. And since everyone cares most about themselves, you should start by understanding how each decision-maker thinks about personal risk. And since personal risks are emotional, be prepared for seemingly irrational behavior when asking for funding.

Several examples of this include:

People hate losing hard-won gains. It's not that executives don't see potential in your idea, it's that they know and fear the potential loss of the investment. There's a psychology to loss aversion. People don't want to lose money they've already gained—even if the odds favor gaining more money in the future. This frustrates innovators because it seems that executives are blind to obvious opportunities. A savvy innovator understands that executives are not blind to opportunities, but are cautious because their reputations are on the line. They're only doing their job—protect the corporate coffers and themselves.

People rationalize risk if they desire the outcome. Individuals will justify all sorts of risky decisions if they want something bad enough—from purchasing a car they can't afford to learning how to sky dive. Same thing in the corporate world. Executives will mentally minimize risks if they like an opportunity until they can rationalize the investment. Many companies have pet projects floating around where some people are shaking their heads wondering if they're too risky (remember the Opportunity Tornado environment from Chapter 3). A savvy innovator makes an idea irresistibly exciting and desirable when asking for funding thus prompting the decision-makers to accept the risks.

Everybody sandbags. Protecting one's personal interests is known as *sandbagging*. You're sandbagging when you skew information such as forecasts and schedules toward your personal interests. Despite all the processes companies put in place to make rational decisions, people still manipulate outcomes to optimize their interests. It's human nature. Sales people underestimate forecasts. Engineers overestimate development times. And executives have been known to manipulate stock option grant dates.

When seeking information from others, *always* assume they are sandbagging. Likewise, when you present an opportunity proposal, executives will assume that *you* are sandbagging. This is why they will question your forecasts, costs, expenses, and schedules. How do they know you haven't manipulated the data or cherry-picked information to get funding? To overcome this, a savvy innovator will have factual data and stay objective.

MY OWN SANDBAGGING STORY

While at Motorola, I started as a product manager for a new digital satellite receiver. I wanted to get the product developed and obtain a large marketing budget, so I presented the highest sales forecast I could get away with. It worked and we shipped the product. Two years later, I became a commissioned sales manager where the same product was part of my commission plan. I then lowered the forecasts so we could hit sales targets and I could get my bonus. Two years later, I became the general manager of a product group with responsibility for the profit of that same product. Now as an executive, I had to reach certain profit margins. So to forecast lower margins, I convinced my management team that competition was driving down industry prices. In each case, I optimized the flow of information for my personal benefit.

Everyone has their own risk baggage. We're victims of our backgrounds and experiences. This is why people say, "We tried that already and it didn't work!" It's typical to point to failure as justification to say no to new ideas. For example, if your company develops products for equipment manufacturers and you want to develop a consumer product, you might hear, "Our company tried launching a consumer product last year. The inventory expenses killed us!" Executives are valued for their experience. Often their learned lessons prevent future mistakes. But past failures also become powerful baggage for rejecting new ideas. A savvy innovator must research an executive's past and the history of any failures. Then use that information to help the executive overcome the reasons and remove any fear of repeating the past.

To get funding for your idea, risk must be managed at the personal level as much as at the business level. Next we'll explore business risk factors. However, while these factors are important, it's more important to know which of these factors lead to the personal risks discussed above and how you'll get executives to accept the risks.

Four Categories Of Business Risk For New Opportunities

Business risk is based on the fact that the future is uncertain and un-expected factors can impact the success of a new opportunity. If we could predict all future events and develop a flawless plan against them, then there'd be no business risk.

There are two primary dimensions of uncertainty that create a business risk for new opportunities: 1) External Uncertainty, and 2) Internal Uncertainty

External uncertainty happens when indeterminate influences exist outside the company such as: uncertain market adoption rates, competitor strategies, or situation factors. Internal uncertainty comes from unpredictable challenges inside the company such as: the unknown ability to develop technology, the lack of key skills, or wavering internal politics. Note that risk is not created by strong competitors or not having critical skills. Risk is created by the uncertainty of what competitors will do in the future and the uncertainty of your ability to build new skills.

The external and internal uncertainty of new ideas can be graphed as two dimensions on a grid to create the four categories of business risk seen in Figure 6.1. Each dimension combines to create a range of low to high uncertainty. The higher the uncertainty, the higher the risk of having success with the new opportunity.

Low Risk Opportunities

At the low end of the risk spectrum is the quadrant titled Low Risk. This is when all of the business risk factors are well understood and easily predicted. For example, your ideas exist in a stable market with known competition, you have customers ready to purchase, and you have a clear track record of developing and selling similar products. This is the realm of incremental innovation and where decision-makers are most comfortable. If executives can achieve desired growth (and bonuses) from opportunities in this quadrant, they don't need to enter other quadrants. And you probably don't need a 30-Day plan to evaluate the opportunity. However, most BIG ideas are not low risk and have many risk factors that must be addressed.

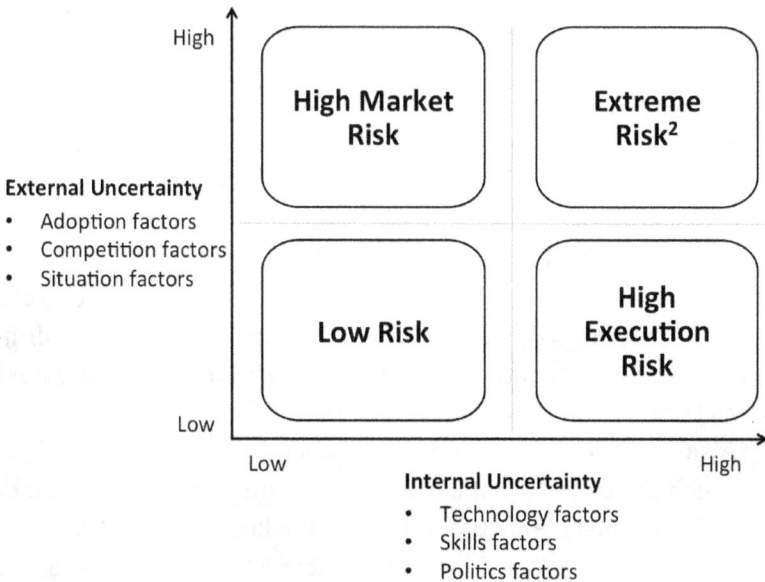

Figure 6.1: Types of opportunity risk

High Market Risk Opportunities

When most of the uncertainty is created by external factors, the business risks reside in the High Market Risk quadrant. Several external factors contribute the most uncertainty including:

   Adoption Factors. Will the new product or service be adopted by customers or not? Adoption factors may impact the success of a new product when you're taking market share from strong incumbents or creating a completely new category of product and must convince potential customers they need it. It may seem difficult to minimize adoption uncertainty, but it's actually one of the easier risks to mitigate if you have the right techniques to understand customer needs and test new concepts. I've seen many companies that could have easily reduced adoption uncertainty well in advance of large investments, but waited too long to ask real customers if they'd actually adopt the new product (the results of Webvan and the IBM PC Jr. come to mind). A savvy innovator lowers this uncertainty by having a customer insight plan at each stage of development to understand customer needs and to determine the concept's value for that customer.

Competition Factors. One of the most common sources of uncertainty comes from current and future competitors. It's easy to list the visible competitors. They currently offer or have announced competing products. But it's difficult to anticipate what the competitive landscape will look like in two or three years. The one thing you can assume is that the competition will not be idle. Consequently for really new ideas where there are no direct competitors in a market, you must address non-competition. A savvy innovator goes beyond general statements like, "We'll watch the competition closely," to a more specific plan that describes which competitors must be tracked, what movements to look for, and how to respond. If there are no competitors, then it must be proven how to convince customers they need the proposed concept.

Situation Factors. This set of factors are the hardest to understand and control. They include uncertainties in the overall economic climate, trends in industry investment, government regulations, and other factors that impact the market potential for a new opportunity. Timing is probably the most critical of the situation factors to clarify for every business opportunity. A savvy innovator is able to clearly answer the question, "Why should we invest in this opportunity now?"

High Execution Risk Opportunities

When most of the uncertainty is created by internal company factors, the business risks reside in the High Execution Risk quadrant. Execution risk can be just as detrimental to the success of a new opportunity as market risk. Markets are littered with great ideas that came out at the right time but failed because of poor execution. The Apple Newton, Google Wave, the Sony Mylo, and Microsoft's Vista are just a few examples. The internal risk factors that contribute to execution risk include:

Technology Factors. These occur with product development and creating new technology. At the high end is when companies try to invent breakthrough technology. Radical or disruption innovation can have large payoffs, but come with a very high uncertainty. Usually only companies with a large R&D budget take on this type of risk. Executives struggle with this. They want breakthroughs (and the payoffs), but they also want the certainty of resource levels and schedules attached to them. A more common technology risk is when a company applies proven technology to existing products or to create new ones. This risk is more about resources and the uncertainty of meeting development schedules due to the number of available engineers, access to the right technical expertise, or unforeseen problems. In either case, innovators in R&D often get frustrated when executives ask for specific dates and levels of required resources. A savvy innovator understands that development efforts are notoriously difficult to predict. Focus on educating executives about the uncertainties and discuss in detail how to reduce this risk to their satisfaction.

Skills Factors. Somewhat related to technology factors, but broader in scope, are skills factors. They're created by the uncertainty of your company's ability and skill to execute necessary tactics. For example, you've identified a valuable product in a growing market but don't have the skills to adequately sell, market, or produce it. This might seem easy to mitigate, but is often dramatically underestimated because we sometimes 'don't know what we don't know.' We think we're able to do what we need to do or think our team can quickly learn. But most of us assume we can accomplish more, learn faster, hirer better and do anything quicker than we really can. However, executives know from experience that anything new is not easy. A savvy innovator is cautious about saying how "easy" something will be. Focus on how to obtain the expertise quickly, not necessarily how to learn it.

Political Factors. Often the biggest threat to real innovation in many companies. We've already addressed many of the political factors that create uncertainty, such as how different types of environments build hurdles for innovators and how executives may attempt to thwart an opportunity. An idea's success or failure often comes down to the company's commitment to resources for that idea, strength in leading the idea, or giving a team the freedom to execute the idea.

Never underestimate the danger of political risks. Success usually requires support at the highest levels of management, and even then success is not assured. A savvy innovator, even in the most progressive innovation environments, is constantly aware of political risk factors and tries to manage them as much as possible.

Extreme Risk

If you're proposing an idea that has high degrees of both external and internal uncertainty, your opportunity is in the category of Extreme Risk. If your company has a limited innovation budget and you're proposing an idea that is very high in both dimensions, you'll have little chance of getting support. It's just too risky for most executives. However, this is where radical and disruptive innovations often live. Your company just might be looking for these high-risk, high-potential opportunities. If an idea falls into the extreme risk category, a savvy innovator is especially diligent about understanding and addressing the most critical risk factors that concern executives.

No one expects you to eliminate the uncertainty related to your idea, but executives do expect you to know the factors that will most impact the success of your opportunity and to have an intelligent response ready when questions come up—and they always will.

Lower Risk With A Minimum Reasonable Ask

A key strategy for lowering risk is to ask for the least amount of investment to move the idea forward. This level of resources is called the Minimum Reasonable Ask (MRA). It includes the people, money, time, and other resources necessary to reach a short-term milestone. That milestone must be clearly defined to provide more data that will further prove the validity of your idea and remove the uncertainty. If your idea has high degrees of uncertainty in any dimension, an MRA is critical when asking for first-stage funding.

How to use an MRA

Say you're presenting a big opportunity that you estimate will generate over $200 million of new revenue over five years but requires $30 million of investment before seeing the first dime of profit. Seems like a great investment, but the potential loss of $30 million feels overwhelming. This is where the MRA comes in. At the executive inquisition you say, "This is a big opportunity for us. We can further validate this opportunity by learning more about the technology and the market. I recommend we invest $175,000 to develop a rapid prototype and interview customers. This should give us data to make better decisions." An MRA of $175,000 allows executives to minimize the potential loss of the investment until you learn more and require larger funds.

HECTOR SUCCESSFULLY USES AN MRA

Hector is the CTO for a large defense contractor and has asked his CEO for $5 million to develop an idea for new technology that spawn a range of new products. He believes the investment will provide a ten-fold return for the compa-

---

ny, but his request was denied. He asked a savvy innovator, Sandra (a senior VP of marketing) for some advice. Sandra assured him it was possible to go back with a different proposal, but he'd first need to examine why his idea was turned down and figure out what information his CEO needed to take a risk on his project. Specifically, Sandra advised:

Ask where the $5 million would come from: She explained that innovators must understand there are other programs and budgets that may already be on the table. Innovators must be able to answer, "Why should we fund your idea as compared to the all the others being proposed?"

Develop an MRA: Sandra explained that innovation projects must have multiple stages. Asking for resources that only fund the first stage of a project, based on clear milestones, lowers the risk for executives. Hector could then work to reach those milestones and go back with stronger data to get support for additional funding.

Understand the CEO's real objections: Sandra encouraged Hector to ask the CEO directly, "Please help me understand your exact concerns so that I can provide the information you need." Don't assume you already know the answer.

Hector took Sandra's advice and asked for a second meeting with his CEO. He explored the CEO's concerns and the impact that a $5 million investment might have on the company. Hector received the approval to move forward. They agreed on an MRA of $500,000 to prove the concept with more funding available if Hector hit his milestone with positive results.

SAVVY IDEA: CONSIDER THE LEXUS
IN THE PARKING LOT

Remember that concerns of personal risk are based on human needs—happy families, financial security, a self-realized life, a nice home, etc. As I was discussing this concept one day, I saw outside in the parking lot a perfect symbol—a black Lexus SUV. It represents a level of achievement that people attain through hard work and perseverance. It's also a symbol of calculated risk. A Lexus is not flashy, but is stable, practical, and dependable with just a touch of opulence. Once someone has a Lexus, they certainly don't want to lose it. Other symbols are similar—private school, a retirement home, or world cruise. Whatever it is, any investment decision that risks an executive's status, financial well-being, or dreams is going to be met with fierce resistance. A savvy innovator understands what's at risk for decision-makers and makes sure that it's safe. Whenever discussing or presenting a new opportunity with an executive, always consider the Lexus in the parking lot.

The next chapter addresses another major challenge that every innovator will face: how to effectively determine and communicate the value that your idea has for customers.

CHAPTER 6 REVIEW—THE RISK CHALLENGE

Risks are a critical component of any executive's investment decision. As a savvy innovator, you must address the risk factors that each executive cares about most and provide the data that makes the risk acceptable. You can do this by taking a systematic approach to break down risk into its underlying factors, identify and address each decision maker's concerns, and use a Minimum Reasonable Ask to lower risk and make it easier to fund your idea.

Five Savvy Success Strategies

As you hone your skills, practice these five strategies to understand and mitigate the inherent risk in your idea.

1. **Address each executive's critical risk factors as they relate to new opportunities.** Every executive will have unique personal concerns. You should never assume you know what these concerns are. The only way to know what risks an executive will accept is to ask. It may not be possible to address every concern, but showing a willingness to try will build respect and support.

2. **Avoid general risk statements.** Claims like: 'competitors will enter the market,' 'prices will come down,' 'development schedules might be missed,' are meaningless because they don't specify what's creating the risk, the potential impact on the success of the opportunity, or most importantly, detail a plan to minimize the risk. You don't need pages of risk statements in an opportunity proposal, but you do need to identify the major risks and how you plan to systematically reduce each one by getting better data through each stage of opportunity development.

3. **Get agreement on your plans to learn more.** The only real way to lower risk is by learning and obtaining more accurate data. If you're developing a new technology while isolated in a lab or entering a new market with little customer knowledge,

Dorian Simpson

the risk of wasting your investment is high. To get executives to accept risk factors, first identify those factors then get agreement on your plan to learn more. Your plan might include customer research, finding a technology expert, or any activity that you mutually agree will remove uncertainty.

4. **Understand your funding quotient.** Executives will always compare your funding request relative to the amount of funds available. If your idea takes 50% of the funds, then your odds of getting it are low. Calculate a simple quotient by dividing the estimated level of investment you're seeking by the available funds. For example, if you require $2 million and the innovation fund is $10 million, then you are asking for 20% of available funds. Is this acceptable? Maybe. It depends on your company's portfolio goals, the potential reward, and number of ideas executives want to simultaneously investigate. Developing an MRA is one way to lower your relative funding quotient so that it's easier for executives to say yes until the full level of investment becomes feasible.

5. **Gain support from subject matter experts.** Executives may be more likely to accept key risk factors if your opportunity proposal has approval by a respected subject matter expert. Get a senior marketing colleague to review marketing elements, a financial colleague to review your financials, etc. They may not agree with every aspect of your proposal, but they should be able to provide feedback, guide you on developing a plan, and coach you on how to communicate risk factors when you face the executive inquisition.

BOB AND NANCY (CONTINUED)
Bob understands the real risks.

Bob received another chance to develop an opportunity proposal for the HM-3000 wireless heart monitor. To guide Bob, Nancy sent him an email with a list of questions that she needed to have answered. To make sure he gets it right this time, Bob called Keith, the market analyst who Nancy was preparing to take over the project.

Bob: *Hi Keith, this is Bob in engineering. I could really use your help. I know our paths don't cross much, but I'm looking at a new product opportunity and wanted to pick your brain. Do you have an hour in the next day or so?*

Keith: (Uh…analyzing new opportunities is my territory. What is Bob doing?) *Hey Bob. It has been awhile. Sure…what are you working on?*

Bob gives Keith an overview of the HM-3000, and taking a lesson from his talks with Nancy, skips the technical details for now.

Keith: (Hmm, I should at least see what he's up to.) *Sounds interesting. How's next Thursday afternoon.*

Bob: *I know you're busy, but this is kind of urgent. Nancy has given me a deadline to develop a proposal. She said you'd be a great resource, and I was hoping you could give me some guidance. How about over drinks? I'm buying.*

Keith: *OK. Let's meet after work tomorrow. There's a pub just across the street.*

Bob: *Great! I know the place. I'll see you there.*

At 5:00 the next day, Bob and Keith meet at the pub. After the usual chitchat, Bob gets to the point.

Bob: *When I met with Nancy, all she seemed to want to talk* *about were the numbers. I tried to pull these together, but they didn't seem to satisfy her. What do you think?*

Keith: *Well, you need to know Nancy. I've probably presented 20 opportunities and she's never accepted an idea on the first round. That's part of her style. She wants more data and believes you can do a better job if you go back and do more research.*

Bob: (That's a relief…it's not just me.) *She also really cares about risk. She used the term a lot. But there's always risk with a new opportunity. What exactly is she looking for?"*

Keith: *That's a tough one. Risk is kind of like art; you know it when you see it. With Nancy, if she doesn't understand something, she'll keep talking about risk. One thing to know is that her last big project before she came to MedCo failed miserably. She's more cautious now and wants to know everything. She focuses a lot on getting customer orders before the product is even launched. That's what caused her previous project to fail. They didn't even try selling until the product was ready. By then it was too late for them to fix some big problems.*

Bob: (Now Nancy's questions are making sense.) *I had no idea. What else does she like to poke at?*

They continue to clarify Nancy's expectations as Keith shared more about the types of risk Bob should address and how Nancy likes to see information presented. Keith offers to review Bob's plan before he presents next, but can't devote time to actually doing analysis, particularly when the project has not had any formal approvals.

Keith: *If you want to really impress Nancy, you need to get some real customer data. Go see Jenny over in the consumer products group.*

At Keith's suggestion to talk with Jenny, Bob wonders what she could possibly offer and why he'd need any market research…this is not a consumer product, he thinks to himself.

WHAT'S GOING ON?

Bob is starting to understand what risk really means to Nancy. He's also realizing that Nancy is a real person who has personal needs at stake that he must satisfy. He's starting to think like a savvy innovator.

Will Bob be able to figure out how Jenny can contribute to his progress?

(To be continued…)

Chapter 7

THE CUSTOMER VALUE CHALLENGE

Using Customer Insight To Prove Value

A rock solid opportunity proposal shows how your idea creates value for customers—as proven by real potential customers.

Rate Your Current Skills

- Who is your typical customer and how clearly can you articulate their problems and needs?

- How exactly does your idea fulfill customers' specific needs, wants, and desires?

- How would you prioritize your customer needs to objectively evaluate the value your idea creates for customers?

- How would you describe your idea's customer value so that executives will absolutely agree with you?

- What is the best research technique to gain customer insight for concepts that don't currently exist?

Introduction

Remember that your internal customers for your ideas are corporate decision-makers. For them to accept the risk of investing in your idea, they must first be convinced that real customers will actually buy (or adopt) the end product or service. If you can prove real people would say, "Wow! That concept is cool. I'll buy it!" then your executives will be eager to hear the glorious details of how you will convert that customer excitement into profit. This starts by proving your idea has customer value.

As defined in Chapter 5, customer value equals how much worth a customer places on your idea compared to buying something else or doing nothing. A savvy innovator builds a solid case to show customer value by defining the potential customers and answering why the idea has value for them and how much value the idea creates. This will be the basis for every analytical and strategic element that follows. The challenge is getting this information quickly, efficiently, and accurately—based on real customer input—when all that exists at this point is a concept.

Determining What A Customer Values

Many innovators struggle with proving real customer value since they either: 1) believe the value is obvious, or 2) may not know how to get actual evidence. Either of these makes costly assumptions and can prevent the innovator from taking time to find the supporting data. Many innovators claim with exuberance, "This idea is so amazing, customers will love it! It uses this technology and has these features…!" The excitement is great, but decision-makers hear blah, blah, blah since there's no hard data to make a funding decision.

Customer value doesn't exist in the features and functions of products or technology; it's in the minds of customers. Put simply, value only exists if your customer says it exists. A savvy innovator clearly defines the potential customers' needs and then proves the innovation meets those needs better than any alternatives. You don't get many chances to present an idea. A sure way to turn off executives is to create something "amazing" that isn't backed with data to prove customers will find it amazing as well.

NATHAN STRUGGLES TO EXPLAIN VALUE

Nathan worked for a software company developing 3D video editing tools for professionals. He was presenting a new 3D tool for non-professionals at a funding meeting. He started by showing drawings of the concept. Nathan was excited and had spent a lot of time making the screen shots look professional. The CMO asked the first question after two minutes, "I've seen other companies doing this. How does this compare with their applications?" Nathan responded, "Yes, there are several startups, and we're pretty sure some of our key competitors are also working on 3D tools, but they don't have our interactive website that makes the editing easier for non-professionals." The CMO asked, "So what's really unique here? I know the other websites are interactive. What will drive demand from customers?" Nathan responded, "Well, I've identified a list of features the other applications don't have." The CMO pressed, "Ok, but how do you know this is what customers want?" Nathan paused. He wasn't sure how to answer. He thought, *Who was he expected to talk to so soon?* He finally said, "Our tools will be more advanced than anything on the market." This wasn't what the CMO wanted to hear, but he stopped his questions and let Nathan continue. Nathan went on to describe the potential price, revenue and ROI. The meeting ended quietly as the executive team agreed to "think about it." Nathan was doomed. He completely missed the CMO's need to see data that showed what potential customers needed in a 3D editing product that would make them choose Nathan's product over all others. Without this data, they found nothing else credible and tuned out the rest of the presentation.

The Customer Value Equation

Proving whether a concept has customer value starts by understanding how people make purchase decisions. Remember how you bought your last TV. You may have visited a retail store or searched online and considered the screen size, features, price, customer reviews, brand, and design and thought, *This one*. Your brain weighed the various attributes of each product and then made tradeoffs based on your needs and the attributes that best fulfilled those needs. These tradeoffs can be summarized in the Customer Value Equation.

Customer Value = Fulfilled Needs - Pain

To solve this equation, you must understand how well your idea fulfills a customer's needs relative to the "pain" of adopting your idea. The more needs your product or service fulfills, the more value it has and the more pain the customer can tolerate. If alternatives exist, and there almost always are, then you must also show that you're creating more value than those alternatives.

Of course the definition of 'need' is very broad. As a consumer, I don't *need* a 5000 square foot home. As a small business, I don't *need* a 100Mb/s web connection. Need is a perception in the mind of customers that includes anything a customer wants, desires—and in some cases—actually needs. There are financial needs, the need to complete a specific task, or the need to feel proud about a purchase. In short, *customer need* describes anything that a customer *believes* is important when considering adopting or purchasing a product or service. The definition of *pain* is also broad. It includes the pain of spending money, the pain of learning something new, the pain of driving around to find a dealer, or figuring out how to discard whatever product your concept is replacing. A savvy innovator explains the type of need that the BIG idea is fulfilling, how important that need is to potential customers, and whether those customers believe the idea is superior to alternatives.

The Customer Needs Priority Pyramid

Customers typically have many needs, especially when buying something complex like a car, home, or electron-tunneling microscope. Because customers have many needs, you must determine which ones are the most important. Some needs must be met before customers will even consider buying your product or service. Other needs may have value, but are less important. By prioritizing customer needs in a hierarchy, you can determine exactly how and why your idea is creating value. This hierarchy forms a structure called the Customer Needs Pyramid (CNP). At the base of the CNP are the most critical customer needs. The importance of those needs decline as you move up the hierarchy. Your challenge is to uncover these needs, prioritize them according to the CNP, then explain how your idea fulfills needs that aren't currently being fulfilled by other products and services.

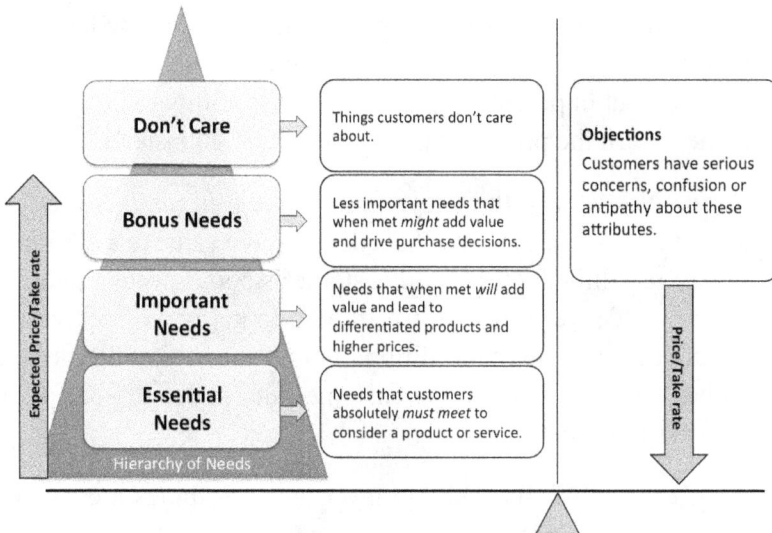

Figure 7.1: Customer needs pyramid

The four levels of the CNP are shown in Figure 7.1. Each level of the pyramid might include a variety of financial, functional, or psychological needs. When seeking the exact same type of product, some customers will have critical needs to accomplish an explicit task or solve a specific problem. Others might have more critical psychological or financial needs. I know many people that would never buy a Hyundai because they think it would "damage" their image (a psychological need)—regardless of the Hyundai's price, features, service, or any other factors. The four levels of a CNP include:

Essential Needs. These form the base of the pyramid and must be met by your product or service or customers will never consider buying it. For example, if someone is purchasing a tripod for a camera, essential needs might include: it must be lightweight, it must attach easily to the camera, it must cost less than $30, and it must have positive customer reviews. Once these essential needs are met, the customer will consider fulfilling other needs.

Non-essential Needs. The needs are not essential, but still important. Our tripod customer's non-essential needs might include: fitting in the glove compartment, extending the height to look over objects, or being stable on uneven surfaces. Other non-essential but important needs might be the ability to buy it from home, return the product easily, and spend $20 instead of $30.

Bonus Needs. The third layer contains the needs that are less important than non-essential needs, but have value to a customer if they're fulfilled. Our tripod customer's bonus needs might include: the desire to return it if it fails, to get photography tips, or to recognize the brand. Filling bonus needs may not drive a higher price, but can sway a customer into selecting your product over a competitor's.

'Don't Care' Needs. The top layer includes things a customer doesn't need or even really care about, but often they're part of the product or service anyway thus adding to the cost or complexity. Our tripod customer may not need to store the tripod in a bag or care about a 5-year warranty. It's as important to know what customers don't need as it is to know what they do. By adding attributes the customer doesn't care about, there's no added value and you may be actually reducing value by inflating price or making the product too complex. If a product or service includes things that don't fulfill the real needs, customers will choose lower price options or decide, "I don't need all of that. It will just be more confusing or cause it to break."

Savvy innovators know this hierarchy of customer needs is at the core of good innovation. Most important is understanding that a CNP is based on what each customer cares about most. When you've identified a set of customers that have a similar set of needs, you've defined a potential target market. The more customers you have with real needs you can satisfy, the bigger the market potential is for your idea.

Customer Needs Are Not Features

Separating a 'need' from a 'feature' is one of the most difficult challenges for anyone. Even customers have a hard time distinguishing needs from features. We're victims of what we already know, so we interchangeably mix what we think we "need" with familiar features, functions, brands, and prices. Customers may say they need a tripod to be aluminum when they really mean lightweight. Or they may need it to have independently adjustable legs, but they really mean stable on uneven surfaces. You must go beyond these stated feature requests and uncover real customer needs.

Customer needs are also very much based on market influences. Our neighbors, family and, of course, advertising all combine to make us perceive certain needs. I didn't know, for instance, that I needed to watch movies with earth-shattering bass, to make a meal in less than 60 seconds, or to have a dazzling white smile, but through media and social influences, these all became my needs. The source of customer needs doesn't really matter. But, if you claim, "Customers don't know they need this...yet," you better be prepared to explain how you're going to convince them they will.

Customer needs aren't static either. As markets mature, types of customer needs and their priorities mature with them. I didn't use to need voice recognition on my mobile phone, but now I do. Products and services compete by fulfilling needs at different levels. But since customer needs change as a market matures, to stay competitive, innovators must be aware of market trends, consistently evaluate the latest needs, forecast the direction they are going, and continue to innovate.

Disruption Happens When Substantially New Value Is Created

When an innovation dramatically meets important essential needs or removes significant pain better than existing products, it has a significant value advantage and disruption can occur. Because price and customer value are highly related, disruptive innovation is a new product or service that meets essential needs at a much lower price or can offer dramatically better performance at the same price. Most people see technology as disruptive innovation, like when transistors knocked out vacuum tubes or DVDs killed VHS. But innovative business models can be disruptive as well like McDonald's use of franchising, Nike's use of sport celebrities, and Walmart's use of inventory management systems.

Use The Customer Value Pyramid For Everything

Once you understand the CNP for your target market, you can leverage it for all further analysis and develop the rest of your opportunity proposal. It becomes the basis for comparing your idea against current competitors and non-competitors, evaluating your innovation's market potential, estimating price, and developing marketing and R&D plans. Say you have an idea for a new phone and want to enter the mobile phone market. To even begin to compete in that market, you'd need to totally understand and meet the essential needs that customers believe they have with touch screens, popular apps, advanced sensors, etc. With your pitch, you'd need to show significantly more value than consumers already get. At this time, Microsoft, Nokia, and Palm are all struggling to create new value for customers in the mobile phone market. Microsoft has had little traction with

Windows-based phones because there's no noteworthy value at any level of a CNP. However, Xiaomi, a Chinese mobile phone manufacturer is having great success by fulfilling essential needs with its Android-based devices that include important high-end features at significantly lower prices. And Google is trying to develop a new way to meet the essential needs of communicating and accessing applications with their Glass device.

A COMPANY STRUGGLES WITH CUSTOMER NEEDS

Josh was the VP of Product Development for a new cloud-based music management application. He had a clear vision for an exciting new service, but things weren't going well. He kept trying to convince his CEO, Marie, that the software was ready and it should be launched into the market. But as Marie tried using the service, she found it slow and confusing. Josh said it was good enough to release and would get better over time. "That's the nature of innovation," he'd argue. Marie wasn't convinced. She kept thinking, "How can I release a service that we're touting as a major new innovation that I can barely use myself?" So Marie asked Tina in her market research team to look into the problem. Tina spent the next week testing the application with a set of target customers. She asked for feedback and summarized the results. Tina reported, "Customers are excited about the service. They love the interface, but it has several key problems that need fixing. There are issues with download speeds and how the service pulls together their music and videos." Josh scrambled to save his vision, but it was too late. Marie put a halt to the project until the problems were fixed.

The root of this problem is simple. Josh directed his software team to develop a whiz-bang interface with lots of cool features. But customers made it clear that the product didn't even meet their essential needs—those at the base of the CNP. R&D could have easily fixed the issues and gotten a valuable product into the market. Instead, they focused on non-important attributes that taxed the available resources and eventually failed by being too late with the wrong product.

Extracting Needs From The Mind Of Customers

To solve the customer value equation for your target customer, you must get inside the mind of potential customers. The tools and skills for doing this are usually categorized as customer research. I've discovered two things about customer research:

- Many people perceive that it's almost useless for innovation efforts.

- If you don't conduct market research, then your chance for success is very low.

How do you to reconcile these two seemingly opposite findings? The problem is that most traditional customer research doesn't work for innovation initiatives. It's tempting for some innovators to point to Steve Job's famous statement, "Apple doesn't do market research." But Apple does conduct research; just not with typical research tools such as surveys and focus groups. They use more innovation-appropriate methods like ethnography, rapid prototyping, and customer trials. Like with Apple, a savvy innovator uses the right type of customer research for innovation efforts.

Don't Expect Customers To Tell You The Future

There is a great debate as to whether customers can tell you anything that leads to real innovation. Let me resolve this debate here—it depends on what you're looking for. If you want to improve current products, then asking customers exactly what they want can help immensely. However, if you're seeking valuable innovation opportunities, you must be aware of *backward-looking insight*. Customers are very good at telling you what they want based on their past. For instance, if your current product is word processing software and you ask customers what they want, they'll give you all kinds of feedback on how to make your product easier along with plenty of ideas for new features. Great if you're looking for incremental innovation. But most companies want innovation based on the future, not the past. Obviously customers can't tell you what they want in the future since they have no awareness of what you can do for them going forward. Did any of us know we could cook a hotdog in 30 seconds, take digital pictures, or make a Facebook page?

Customers can't predict the future, but they can tell you what they want to accomplish and what is important to them in the process. A popular saying is, "Customers don't care about drills, they care about perfect holes." I agree with this and take it a step further to say, "They don't even want perfect holes, they want to accomplish whatever they needed the hole to do for them." If you have a new type of drill, a woodworker can tell you all about the type of furniture he's trying to build and the problems he experiences when making joints or combining pieces. Maybe he doesn't even need a hole.

The Innovation Market Research Challenge

To summarize the challenge of using market research for innovation efforts, let's look at the three most common tools of research:

Customer Surveys. Surveys don't uncover needs or provide deep insight into how customers think. If you ask the question, "What would you like to see in the next product?" and then ask them to select the top five from a list of features, you're conducting backward-looking insight. Yes, you get more statistics (Wow! 32% of customers want an ice dispenser on their refrigerators?) But these results do little to understand true customer needs that lead to real innovation.

Focus Groups. In a focus group you invite 8 to 10 participants to discuss a problem or concept. They work fine if you need validation. However, in a typical 90-minute focus group, where the moderator speaks for 30 minutes, you maybe get six minutes of feedback from each person. This limited time hardly gives you the depth required to understand their needs and explore their attitudes toward new concepts.

Customer Interviews. This is a chance to have a one-on-one discussion with potential customers about problems, needs, and concepts. Granted there are more sophisticated market research methods, but a well-conducted customer interview is the fastest and most efficient technique to gain the most amount of information in a short time period. It's by far the preferred choice for innovation efforts.

However, there is one caveat: if interviews are not properly executed and the results are not properly communicated, then executives won't believe your results. Even if you interview 20 customers who all love your concept, many executives will still be skeptical. They think maybe you interviewed the wrong customers, asked the wrong questions, or skewed the results in your favor (remember sandbagging?). A savvy innovator will take a structured approach to conduct high-quality interviews with the right customers and will communicate the results accurately. In Part III we'll break this process down specifically.

Your goal is not to develop a comprehensive, traditional research plan in 30 days, but instead to gain rapid insight. With no customer research, you risk having a poorly developed concept and opportunity proposal.

SAVVY IDEA: LEVERAGE AS FEW AS FOUR CUSTOMERS FOR YOUR OPPORTUNITY PROPOSAL

You might wonder how you can develop a compelling opportunity proposal that includes solid market research from real customers, in less than 30 days. The short answer is you can't. But you can use a small set of potential customers, conduct short interviews, and use this data to support your initial proposal. You'll be tempted to skip this step. But don't! Remember, you'll always be asked for proof that your idea has value for customers. Without at least some data, you're merely stating your opinion. When time is limited, try having at least four intelligent discussions with potential

---

customers to give you the credibility you need to receive initial funding. Why four? Any less and executives won't believe you, but four shows you made a serious effort. Plus if you interview four customers and three of them provide the insight you need, then you have a strong base to get funding for deeper research with more customers.

In the next chapter, we'll explore how to interpret and use market and technology trends to communicate key elements of your opportunity proposal.

CHAPTER 7 REVIEW—THE CUSTOMER VALUE CHALLENGE

Savvy innovators are able to clearly articulate customer value as the core of their opportunity proposal and use this information as a basis for everything that follows. They determine customer needs to understand how customers make purchase decisions and they conduct customer interviews to develop a Customer Needs Pyramid, which illustrates exactly how their idea has real customer value.

Five Savvy Success Strategies

As you hone your skills, practice these five strategies to determine if potential customers truly value your idea.

1. **Interview potential customers to convince you first.** There are two primary uses for market research: research-to-know and research-to-show. Since you are the first person that needs convincing your idea is worthy of funding, use research-to-know to educate yourself about the value of your concept. Afterwards, use research-to-show to convince others that they should believe you. Over time, you need both, but by starting with research-to-know, you'll gain confidence that will show through in every other aspect of your opportunity proposal.

2. **Gain early agreement on your research approach.** Depending on your environment, interviewing a small set of customers may provide adequate data to support your initial opportunity proposal. However, some executives (especially in Opportunity Clogs and Siphons) may not listen to your proposal without more customer data. Try to get them on board with your market research plan early before going too far in developing your proposal. If executives need to see 20 customer interviews and the results of an online survey before investing, then that's what you'll need to do.

3. **Listen to sales...but not too closely.** As you should never consider yourself the customer, neither should you assume

your sales team members are customers. It's important to get their input on an idea, but they are usually not going to buy your product. Listen to what they say, particularly for their understanding of the competition and how to approach sales forecasts, but use their feedback as input, not answers.

4. **Be wary of the guys down the hall.** It's tempting to ask other colleagues what they think about your idea, but be careful how you use this input. Internal feedback is great if you want input where they have expertise, but they are rarely your target customers and they may be biased one way or another towards your idea.

5. **Partner with your market research team.** These teams may be tasked with everything from customer satisfaction surveys to branding effectiveness. They're a valuable resource for your research efforts. They may have a budget you can use, have access to vendors to find potential customers, or be able to coach you on interview skills. Just be certain not to outsource your research to them at this point. You (not the research group) need to interview the customers to gain first-hand insight into their needs.

BOB AND NANCY (CONTINUED)
Bob takes on the customer value challenge.

Bob is starting to get a clearer picture of what Nancy needs, but he keeps thinking about Keith's last comment, "If you want to really impress Nancy, you need to get some real customer data." What does that mean? Do I need to do a customer survey? I don't have time for that. But again he picks up the phone. This time he calls Jenny, a director of Product Marketing for some of MedCo's consumer products.

Bob: *Hi Jenny, this is Bob over in commercial products engineering. We haven't met, but Keith suggested I speak with you. I'm investigating a new product and hoped you could offer some insight into how to get customer data.*

Jenny: *Hi Bob. Well, I'm not sure what Keith had in mind, but I've helped him in the past to interview customers. Is that what you wanted?*

Bob: (What? I don't even have a product yet, how could I interview customers?) *To be honest, I'm not sure.*

Jenny: (Oh this will be fun!) *OK. I have some time today at 4:00 if you want to discuss it. I'll see how I can help.*

Bob: (I'm too busy for all this, but I have to know what Keith had in mind.) *OK. That would be great. See you then.*

Later at Jenny's office, she cuts to the chase and hands Bob a stack of papers.

Jenny: *OK. Here is the last project I helped Keith with. We do a lot of customer interviews for our consumer products and have developed some guides to help our product managers. I gave these to Keith and coached him on how to conduct an interview. Is that what you're looking for?*

Bob: (I really don't have time for this.) *Well, I'm developing a proposal that I'm supposed to deliver to Nancy in a couple* *of weeks. Keith said having customer data would impress her. I'm not sure what I need. I assume data means some sort of survey. Or maybe you have some data from past reports. I got one from Tanya, but it was really broad.*

Jenny: (I don't have time for this.) *OK. I understand better. When Keith presents a new opportunity, he tries to get a small set of potential customers that he's talked with. I haven't seen what he presents to Nancy, but this must be what he means. Read these documents on interviewing customers. If you need more help, I can set you up with one of the product managers.*

Bob: (Sighs. This is more than I bargained for, but if this is what it takes.) *OK. I'll check it out.*

Bob reviewed the documents and decided he was ready for the challenge of interviewing customers. He followed up with the product manager, who helped Bob understand how to find potential customers and also helped arrange several phone interviews.

Then Bob decided to tackle the revenue forecast. He had done his best with the data he had, but Nancy clearly wasn't buying it. Even though Keith is away in France, Bob calls again, hoping for some more guidance.

Bob: (Man, Keith has a great job.) *Hey Keith. I know you're in Paris, but I really need your help. First, thank you so much for setting me up with Jenny. I've been interviewing potential customers and the input has been amazing. I'm struggling now with developing the financial forecasts. Can you give me any advice?*

Keith: *Bonjour Bob. Glad to hear Jenny could help. You should talk with Chris, MedCo's SVP of Sales and Marketing. He's the master of forecasts and the guy I always have review mine. He's brutal, so if you get by Chris, you'll be fine with Nancy.*

Bob: *That's awesome Keith. Merci beaucoup. I'll let you know how it goes.*

WHAT'S GOING ON?

Bob is starting to think and act like a savvy innovator. After a couple of customer interviews, he started seeing what Keith was trying to tell him. He's not sure yet how Nancy will react to data from only five customers, but Bob's confidence has grown. He's been listening to what the customers have told him about their needs and has already changed his thinking about some of the important elements in the product.

Bob is running out of time. Will he be able to develop forecasts that are acceptable to Nancy?

(To be continued…)

Chapter 8

THE FORECASTING CHALLENGE

What To Do When
There Is Little Data

---

*When it comes to forecasting a new opportunity,
success is all about your data and approach,
not your numbers.*

Rate Your Current Skills

- How convincing is your story with just a small amount of relevant data?

- What is the difference between facts and speculation?

- How well do you understand each type of forecasting variable that contributes to the financial potential of a new opportunity?

- Are your forecasts based on the specific market situation your idea is entering?

- How do you convince executives to accept your forecasts?

Introduction

Forecasting is by far the most controversial facet of any opportunity proposal. Ultimately, your proposal will be judged on the quality of its numbers for a wide range of variables such as the market size, revenue, costs, and expenses. You'll use these numbers as the basis for calculating a return on investment that will become the final gauge for a funding decision. Because personal opinions are so rampant at the early stage of ideas, it's difficult to find solid data, and there is no such thing as a correct forecast. Executives are extremely skeptical about financial estimates early on and will always question, "Can I believe anything about these numbers?"

Only when you remove show-stopping objections and convince executives your forecasts are reasonable will you have the credibility to discuss funding your idea. A savvy innovator satisfies the most ardent critics by presenting forecasts of key variables using a logical approach and as much relevant data as possible.

The Basics Of Forecasting

By definition, *forecast* means an estimation of future trends or events. In the innovation world, everything concerning your new opportunity is a forecast. However, most executives, particularly the most risk-averse ones, need to base funding decisions on solid data. But data, naturally, are relics from the past. The secret is to build a clear and logical bridge between historical data and your vision of the future. If you can't build this bridge, you might not be able to build your case.

AARON'S TROUBLE WITH FORECASTING

Aaron was a product manager for an educational software company. He was confident that virtual reality (VR) headsets were gaining traction with customers and was presenting an idea to develop virtual tours of historic sites. As the meeting started, Aaron handed each executive a VR headset as he

---

explained how VR would open up a whole new market for richer educational and entertainment experiences. He smiled as everyone's faces lit up, experiencing a sample virtual tour of Beijing's Forbidden City. After the demonstration, Aaron pulled up a spreadsheet to present the numbers. "The tourism market to historic sites is a $220 billion global market opportunity. Last year, 14 million people spent an average of $1800 to visit the site you just toured from the comfort of your chair. We expect virtual tours to be 5% of this market in 10 years and we think it's possible to take 10% market share for a total potential revenue of $1.1 billion by 2025." After a bit of silence, Beth, the COO, asked the first question, "Soo…customers are willing to pay $1,800 for a virtual tour?" Aaron paused. He looked at his math briefly. He knew that wasn't possible, but that's what his math said. Aaron backpedaled and attempted to explain that the market would still be big. Unfortunately, it was too late. The excitement in the room turned to confusion as each executive tried to find meaning in the numbers. Aaron was told to come back when he had more reasonable financial projections.

This is an extreme example, but it's typical of many forecasting exercises. High-level numbers are presented, some magic happens, a result pops out, but there's very little logic in the middle. Aaron needed to bridge the gap between the convincingly large size of the tourism market and the realistic potential of virtual tours. While this mistake is often related to market and revenue forecasts, it can easily happen with forecasts of price, costs, development time or expenses.

Failure Is An Option—Poor Logic Is Not

If forecasting the future feels daunting, don't despair. Even seasoned market analysts, investors, and CEOs struggle to develop accurate forecasts. For example, the dot.com phenomenon Webvan (an online grocery delivery service) raised over $1.2 billion in startup funding only to be sold nine months after its launch for pennies on the dollar. One of their primary forecasting bungles was predicting the market size. The founders, industry analysts and investors, all dramatically overestimated the market potential for online grocery delivery. In 1999, one analyst predicted (and many others agreed) that online grocery sales could reach $100 billion by 2009. By 2013 online grocery sales were only hovering around $1 billion. Similar forecasting issues plagued the Segway personal transportation vehicle and the IP phone service Vonage. As of this writing, only time will tell if the high expectations for smart watches, personal health care monitoring, self-driving cars, and virtual reality applications will be met.

You don't need accuracy to get your idea funded, but you do need a logical and convincing approach with credible data sources. As highlighted in Figure 8.1, the best forecasting is the most logical forecasting. You create a convincing path that starts with acceptable data and logically connects through acceptable forecasts of market, revenue, and profit variables. To gain acceptance of financial calculations, a savvy innovator communicates each forecast in a way that decision-makers will think, "This is plausible. The data is reasonable. The approach is rational. The numbers are the right magnitude, and I don't think I could develop a better forecast."

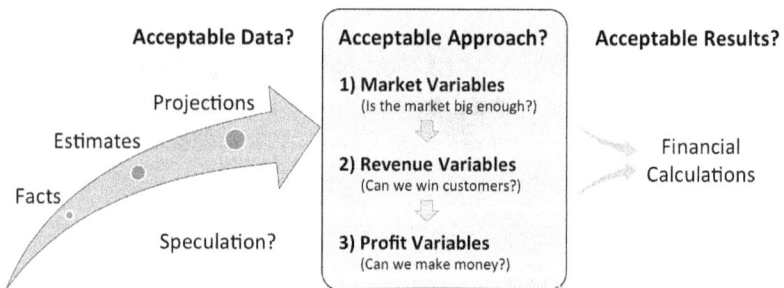

Figure 8.1: The flow from data to financial calculations

Data Comes In A Spectrum Of Truthiness

Remember that *data* is one of those fuzzy innovation terms. In the Savvy Innovator Dictionary, we defined it as: any piece of relevant information used to clarify or validate your opportunity designed to gain support from funding executives. Obviously, the newer and more radical your idea is, the harder it is to find relevant, believable data. Regardless, to create a powerful argument for your opportunity proposal, you must start your forecast with acceptable data. Easier said than done. Accurate data is hard to uncover—there is a range of *truthiness* in every piece of information. You know this, for example, if you've ever tried to analyze the 'internet of things' or figured out which one of your kids broke the TV remote. When developing forecasts for new opportunities, consider these rankings:

Facts. Data from reliable sources using solid methodology can generally be considered facts. Facts could be a history of company and industry sales, information about existing products and companies, government demographics, or other data that has been collected using a reasonable tracking mechanism. This could also be clicks on a website, data collected from a competitor's Bill of Materials (BOM), or even well conducted market research. Any information that is difficult to refute are typically facts. They are the most believable type of data.

Estimates. Once you get past the sources of facts, you start getting into estimates. They're not inherently bad since hard facts related to a specific opportunity are difficult (if not impossible) to find and all innovators must make estimates. In the Nathan story, for example, he should have estimated the number of people who already own VR headsets if a factual number was not available. But warning: if the estimate is not based on facts, it may be seen as a "guesstimate," which rarely has credibility with decision-makers.

Projections. When you extrapolate an estimate into the future, the data becomes a projection. This is what forecasters use in reports like, "The future of healthcare," "The Worldwide Market for Self-driving Automobiles," or "Millimeter-Wave Emerging

Markets." While the forecasters try to base their projections on facts, they're usually stating opinions based on estimates and sometimes even guesstimates. Projections have less credibility than fact-based estimates, since executives often rely on their own projections of the future.

Speculation. Beyond any attempt to rationalize information completely, you step into the world of speculation. Everybody loves to speculate! Speculations on the future of products, technology, and markets are rampant and are constantly espoused by industry executives, reporters, analysts, futurists, technologies, casual bloggers and so on. This category of data typically has no credibility when attempting to justify a new opportunity.

Unfortunately, it's often hard to tell the difference between facts, estimates, projections, and speculation. For example, one innovator I worked with needed to forecast the market potential for a new type of criminal restraint device. To develop his forecasts, he used an accurate number of police stations as critical data. One market report showed there were 6000 police stations in the United States. Was this a fact? As he researched, he discovered it was an estimate based on the number of U.S. cities and an assumption of the number of police stations per capita. As a savvy innovator, he realized this statistic was questionable, so he looked for a more credible fact. He found it from the Bureau of Justice who reported there were 12,501 police stations in the US.

The Garbage-In/Garbage-Out Phenomenon

Many innovators struggle with the familiar problem of *garbage-in/ garbage-out* forecasting. It happens when an innovator begins with a guesstimate (garbage in) of a key variable then piles on assumption after assumption until a number is revealed (garbage out). They may have thought deeply about each variable, but as the layers of assumptions are peeled back, there is nothing left but the inaccuracy of the original guesstimate. These models can take hours to explain, but usually no one ever believes the results.

Every sound argument, be it a criminal legal case or proof of a mathematical formula, must be based on facts or the argument crumbles. Savvy innovators start with relevant, nearly undisputable facts to gain acceptance on their forecasts. Speculating about the future is fine—and expected—but if your forecasts are based solely on projections that aren't rooted in facts, you face a great danger that your whole proposal will be viewed as something that should be quickly disregarded.

Why Real Customer Data Is Critical

For really new ideas, market size and revenue forecasts usually rely on projections based on a loose connection to real facts. Sometimes this is understood and acceptable. Otherwise, the only thing that will be accepted is the data gained from real potential customers. While you don't have time or enough information to conduct statistically relevant research, to be credible your data must prove that real potential customers find your concept valuable. Some executives will need you to take this insight as far as having real customer orders. For business-to-business products, this may be a purchase order. For consumer products, this may be a letter of commitment from distributors. For early-stage ideas, it's tough to get order commitment. However, you must be able to provide some amount of direct customer data to prove that you know who the customers are, that you know how to find them, and that customers readily see value of your new concept.

Creating Acceptable Forecasts

Successful forecasting answers three primary questions as shown in Figure 8.1:

- Is the market big enough?
- Can we win customers?
- Can we make money?

You're probably familiar with the overall approach to answering these questions. You need to estimate the size of the market, estimate the number of purchasing customers, then develop financial estimates by projecting price, costs, and expenses. Conceptually, the steps are easy. And in Part III, we'll walk through this process using a case study. For now, let's address the market situation.

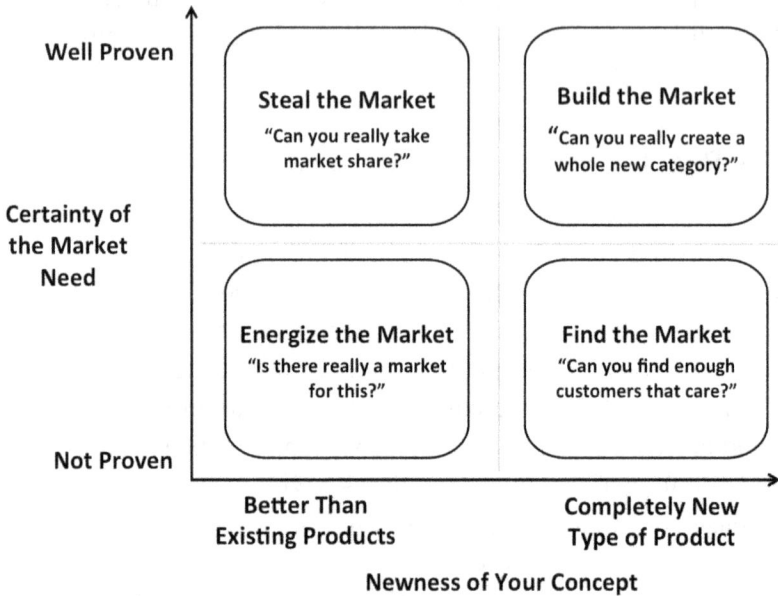

Figure 8.2: Four possible market situations

Forecasting Starts By Understanding The Market Situation

How you forecast each variable and answer the three primary questions varies widely by the market situation your idea is thrust into. As shown in Figure 8.2, new concepts enter into one of four market situations based on two dimensions. The first dimension is the newness of your concept. At one end of this spectrum your idea is just slightly different (hopefully better) than existing products. At the other end of the spectrum your concept is completely new to the world.

The other dimension is the certainty of a market need. At one end of this spectrum are well proven needs, like preserving food or reducing pain. At the other end of the spectrum are uncertain needs, like shaving functionality in a mobile phone (yes, a real product launched in 2009). These two dimensions combine to form four types of market situations.

Stealing the Market. If your concept is entering an existing market with direct competitors, then it must be good enough to steal the market. To convince executives your concept is viable, you must prove to them you truly understand the current and future state of competition. Forecasting will be based heavily on determining exactly what is new and different about your concept compared to alternatives. And not just now, but the advantages you expect to have once it hits the market and how you plan to stay ahead of the competition. Successfully projecting price, costs, and expenses are critical in this situation since existing products are likely maturing. Stealing market share will depend on your ability to compete directly. Any idea that is a "new and improved" version of existing products in a mature market fits this category.

Building the Market. If your concept is really new, and the customer need is clear, then your primary forecasting task is to prove that you can build the market. This is often an ideal market situation assuming that you can really prove customers need your concept. You'll need to estimate what it will take to educate the market, build the sales infrastructure, and create market demand. If the technology is unproven, you'll also need to project the time and resources to develop the technology. Tivo, one of the first digital video recorders, is a great example. Tivo saw a very clear need (and desire) to pause and record television and achieved success by building a new market for very new type of product.

Finding the Market. If your concept is new, but the customer need is uncertain, then you'll have to find a market. This is an extremely common, but difficult situation for innovators. Most executives don't want to risk investing in an idea that does not solve a large, pressing, and proven market need. You'll have to explain the potential of your idea by first identifying the market need that is being solved and then determining if you need to steal the market or build a market.

Energizing the Market. This is when current products exist, but they haven't yet found a lot of demand with customers. You'll have to first and foremost convince executives that you can energize the market. We've all seen the launch and fizzle of products like flavored milk and netbook computers. Say it's 2006 and you're an innovator at a company (other than Apple) and said, "I propose we develop a tablet computer." Your executives might point to the failure of IBM, Dell and others with their tablet computers and say, "That market is dead." You might get the same response to QR Codes (those blocky squares meant to replace barcodes), digital audio tape, or flying automobiles. A savvy innovator has to determine whether the market for these products really exists and whether the other companies just didn't know how to execute or their timing was off or if customers really aren't interested.

The Market Situation Is Not Obvious

Never assume the market need or the perception of newness is obvious. You might think the concept is completely new and fulfills a well-proven need, but as you talk with executives, they might say: "Yeah…but that's already out there!", "Customers won't buy that.", or "Hasn't that already been tried?" Maybe the executives don't understand your concept. Or maybe they're right. Each market situation has unique forecasting challenges. In all cases, you must forecast the market size, revenue, and profit. However, each case raises fundamental questions that must be answered and defended successfully before your request for funding will be taken seriously.

The Right Approach For Each Market Situation

At this point when evaluating a BIG idea, focus on finding data that answers the three questions: "Is the market big enough?", "Can we win customers?" and "Can we make money?" Approach these with as much credibility as possible while understanding you will never have perfect information.

JOAN APPROACHES FORECASTING FOR HER BANANA ENHANCER

Let's say Joan, a product manager for a small kitchen appliances manufacturer, has an idea for a banana life-enhancing appliance. She is aware that bananas are the number one fruit sold in the world. As a banana-lover herself, she believes that if a banana is too hard or too ripe, consumers don't want it. She recently saw an advertisement for a food storage container (like Tupperware™) that helped keep fruit fresh with the aid of a new semi-permeable material. After some research, she learned that bananas release ethylene gasses. Keeping them in a bunch or with other fruit only hastens their ripening. So eureka! She thinks, "What if my company could develop an appliance that could control the rate of ethylene to slow or speed up the ripening of bananas? We could call it the BananaEnhancer."

Joan might expect to hear, "This is crazy. Who is going to pay for an appliance dedicated only to bananas?" So clearly she needs to develop acceptable forecasts as part of her opportunity proposal that answers the following questions:

"Is the market big enough?" Based on industry reports, Joan learns that bananas are the number one fruit consumed in the US with over $1 billion sales in 2010. It's clearly a big market. However, this data is almost meaningless to justify the market size for her device. She needs to go deeper to determine how

---

many customers preserve food and what they are willing to spend. Because of her need to find a market, she'll need to estimate how many consumers care about preserving bananas and what they might pay for a reasonable solution. She'll need to first find the market and then determine how she could build the market profitably.

"Can we win customers?" The answer to this question comes in the form of a revenue forecast that is based on an estimated price and the number of customers expected to purchase the concept over time (a.k.a the take rate). The take rate is by far the most controversial of all forecasting variables. Joan must first estimate a wide range of variables like how many potential customers can actually access her new product, how many of those can be made aware of the product, and ultimately, how many she can convince to buy the product in a specific time frame. These variables are highly dependent on the projected product attributes, price, and quality of marketing activities. Because of the number of variables involved, revenue estimates are "ripe" for a garbage-out forecast (yes, pun intended).

"Can we make money?" If you're able to gain acceptance for your forecast of how many customers you'll win, then you can answer this final question. For Joan to forecast profit, she first needs to make projections for all of the variables related to product costs, R&D expenses and sales and marketing expenses. Combined with price, these variables determine the gross and operating margins of her concept. These margins are critical metrics for most executives to judge the viability of new ideas, so forecasting these variables for new concepts cannot be taken lightly.

Now Joan needs to get executives to accept her answers. To do this, she'll need to understand her executives' attitudes and the level of data she needs to obtain for each answer to be considered "acceptable." Since the BananaEnhancer is an obscure concept, executives will be highly skeptical of its value to customers. She needs to get real data from real customers, prove that she could do a trial of the product, and even prove the technology. If she's entering a crowded market filled with other banana storage devices, she'll need to go deep into the competition, costs, and advantages of her idea versus the competition.

In any case, none of Joan's answers will be totally believable, but she must be able to answer definitively, "At this point of evaluating this opportunity, here are the facts and here is my approach to forecasting."

There are many approaches that companies use to develop forecasts. You might be familiar with some of them such as: the Delphi technique (averaging a range of industry experts), Monte Carlo methods (using computer modeling to estimate the probability of outcomes), Bass Diffusion modeling (attempting to model the market using a range of adoption factors), and scenario analysis (developing multiple forecasts based on a range of assumptions). It's great to use these techniques if your company has adopted them and found them valuable in the past. However, they can be problematic because often you spend too much time explaining the method, which just complicates matters. Usually it's best to take a straightforward approach using a simple profit and loss model to forecast the market size, revenue, and profit. And then…use the following secret to forecasting success.

The Real Secret To Forecasting Success

To solve the problem of executives' skepticism, the real secret is… (pause for effect) make them part of the forecasting process. This may not have surprised you, but it is the part of forecasting that most innovators skip and savvy innovators embrace.

The success of your opportunity hinders on understanding each decision-maker's needs and making them part of the forecasting process. Otherwise, you simply won't be properly prepared to weather the executive inquisition and the objections that come with every forecast. To build support for your market and revenue forecasts, work with the most senior sales or marketing executive you can find. For R&D investments, get support from senior technologists. And once you have your forecasts together, they must all be approved by a respected financial executive for anyone to accept the results. Addressing executives' needs makes it difficult for them to say "no." Getting their participation makes it easy for them to say "yes!"

By gaining their input and participation early on, the chance that they'll accept your forecast when it's time to decide goes up dramatically. But before seeking their feedback, have a forecasting approach in mind that considers the underlying variables contributing to your revenue forecast. Most executives, although busy, are very interested in learning about new opportunities and working directly with innovators. They may start out as your biggest critic, but often end up being your best ally.

SAVVY IDEA: ALWAYS DEVELOP YOUR OWN FACT-BASED FORECASTS

Innovators often rely too heavily on stand-alone industry forecasts. So to add credibility to those forecasts, develop estimates based on your own research of market potential and revenue. For example, one innovator was forecasting the market potential for 3D printer components. One industry report projected that the market for high-end 3D printers would be $560 million dollars by 2020. To validate this, he took another approach by calculating all of the types of companies that would purchase 3D printers (such as machine shops, design studios, and product development companies), how many they might buy over the next few years, and average prices. He concluded that the market would be about $420 million by 2020. While his own estimate was far from perfect, the process of developing it allowed him to think through the market more deeply and gave him more data and more confidence when presenting his forecasts.

In the next chapter, we'll complete the discussion of key financial terms as they relate to innovative ideas and how to develop a comprehensive financial story that executives need to hear.

CHAPTER 8 REVIEW—THE FORECASTING CHALLENGE

Creating acceptable forecasts starts with a clear hypothesis and then finding the best data you can to create a believable story. There's no such thing as a correct forecast, but you can improve your chances of getting initial funding by focusing on a rational approach for each critical variable that executives accept. This will then give you the time and resources you need to conduct more research and analysis for future phases of your opportunity development.

Five Savvy Success Strategies

As you hone your skills, practice these five strategies to develop acceptable forecasts for your idea.

1. **Focus on the data and approach first.** Whatever numbers you present, assume they'll be wrong. You know it. I know it. Executives know it. And because of this, decision-makers will assume you're presenting the most favorable numbers possible to gain support. Whether you are sandbagging or not is immaterial, but you must go with that assumption and ensure they first react favorably to your data and approach. Only then will they accept your numbers.

2. **Be optimistic, but realistic.** While your idea may be the next Google and deliver similar results, if you present wildly optimistic forecasts, you'll come across as a head-in-the-clouds innovator. Your best approach is to offer realistic revenue and profit estimates then let the huge potential upside of the opportunity speak for itself. Executives will always come to their own conclusions as to your idea's home-run potential.

3. **Focus on a persuasive story, not forecasts.** Developing forecasts for a really new concept is more like a criminal investigation than a rigorous analysis. You must pull together as many facts as you can and then weave them into a compelling story. Similar to a good trial lawyer, a savvy innovator can persuade

an audience to agree using just a few relevant facts. When developing forecasts for a new opportunity, you'll rarely have the smoking gun, so build the best argument possible and then test, refine, and test again.

4. **Don't rely on hype.** Innovators can get really excited about their ideas based on market hype. The self-driving vehicle market created by Google's autonomous driving experiments is a good example. It's exciting since these cars will remove congestion, add safety, aide the blind, and have many unforeseen customer benefits. However, excitement alone doesn't provide the business rationale for investing in it. Getting ideas funded requires you to develop numerical forecasts for specific products, markets, and customers.

5. **Leverage experts as much as possible.** Just as with other risk factors, often the only credible place to get acceptable data is from experts who have experience in R&D, sales, and marketing. For really new technology and concepts, you might have to go outside your company to find a credible source that can provide a magnitude of the R&D effort or possibly an expert opinion on whether the product is possible.

BOB AND NANCY (CONTINUED)
Bob tackles revenue forecasting.

Bob is on a roll now. He's been working through the risk factors that Keith helped him with. He can also articulate customer needs and how his idea will meet them. Now to forecast the key variables of his idea…but how? Luckily, Keith suggested calling Chris, MedCo's SVP of Sales and Marketing.

Bob: *Hey Chris, I'm working on trying to forecast the financial potential for a new product I'm investigating. Keith said you're the man. But I have to admit; I have no idea where to start. I know you're busy, but could you coach me through this? I promised Nancy a proposal next week that includes financial forecasts.*

Chris: (Hearing Bob's desperation in his voice.) *OK. Come by in the morning and let's see what you have.*

The next morning, Chris suggests they walk to get coffee. Bob talks about the concept, his discussions with customers, and the data he's found so far. He pulls out his presentation, which showed a detailed schematic and had "Wireless HM-3000" written across the top.

Chris: (Ignoring the schematic.) *First of all, Nancy will never believe your revenue forecast. No one will. She'll want to see the growth in the market category, and she'll need to know that what you are pitching has a chance in the market. Let me see what you have so far.*

Bob: (Showing Chris his forecast.) *This seemed reasonable to me based on the data I have.*

Chris: (Chuckles to himself.) *You have here what I call a "Hail Mary forecast." That's when you take an estimated market size from some random analyst and just assume we'll capture a large percentage of the market. No one believes it,* *and it has no credibility. What else did you show her to justify this?*

Bob: (Looking perplexed.) *Well, let me walk you through the product features.*

Chris: *Hold on. I don't need to see those. You said you did some customer interviews? What did they tell you?*

Bob: *The customers were very supportive. They liked the idea and want to learn more about it. Two of the doctors said they were waiting for something like this.*

Chris: *That's encouraging, but did you get any insight into the challenges to purchase? Or how it might compare to other similar solutions? You need to take a complete market view of this and build a path between zero sales and your revenue estimate. If you can't explain how you'll achieve the revenue forecast, the forecast is meaningless.*

Bob: *But you just said no one will believe the forecast.*

Chris: *True, but if Nancy believes that your product is creating something unique that customers actually want over existing options and you understand how to find customers and convince them to purchase, she'll at least believe there's a chance to hit reasonable sales. It's about telling a complete story more than your forecast. Does that make sense?*

Bob: *I think so. How do I go about doing that?*

Chris goes on to describe the connection between sales channels, marketing activities, and revenue. They discuss the likely hurdles of customer adoption and how Bob might explain how he'll achieve sales.

Bob: (Feeling a sense of relief.) *OK. I think I can take a shot at this now. Can I meet with you again to review my next effort? Keith said you're brutal, and I want to get this right.*

Chris: *I have to run to a meeting. I'd be happy to review your* *proposal. By the way, your financials are a mess. You need to go see Bridget in finance.*

As Bob heads back to his office, he wonders what Chris meant by 'my financials are a mess'. His confidence on this proposal had been growing, but now he's not so sure. The customer interviews took up much of the last two weeks. How can he possibly finish in the next few days? It has ballooned to over 100 presentation slides, and now Chris says to work on the financials.

WHAT'S GOING ON?

Bob is facing the common challenge of developing a defensible financial forecast. He realizes now that there are no right answers. But he must have a good hypothesis, good data, and clear business thinking. Now Bob has to tackle one more innovator challenge if he wants to hold an intelligent conversation with Nancy. He must develop financial information that meets her needs. He's hoping Bridget can help.

Can Bob learn the language of finance?

(To be continued...)

Chapter 9

A 30-MINUTE CORPORATE INNOVATOR MBA

Key Financial Terms And Definitions That Innovators Must Know

---

Savvy innovators don't need a finance degree, but they definitely need to be fluent in finance.

Rate Your Current Skills

- How does your CFO define Return on Investment?
- What is the difference between costs and expenses, and how does income differ from margins?
- What is your company's hurdle rate for making investments in innovation projects?
- How would you describe the components of Cost of Goods Sold?
- What is your response to, "Calculate a five-year NPV using a discount rate of 13% and ignoring residual value"?

Introduction

When I received my MBA, we covered a lot of topics from operations to marketing to ethics, but we mainly focused on financial skills with courses in corporate finance, managerial accounting, and financial accounting. Although my classes on entrepreneurism and marketing were often more fun, my finance classes have been the most valuable to my career since those skills are difficult to obtain without formal training.

If you don't have an MBA, I'm not suggesting you sign up for classes and enjoy all the sleep that I did as my retired accounting professor droned on about debits and credits. But if you want to speak intelligently with financially minded executives, take some time to get steeped in key financial terms and concepts.

In this chapter, we'll cover the elements of financial statements that will always be used to evaluate your ideas. You should know specific terms and common mistakes to avoid as well as how to respond to certain questions you'll hear during the executive inquisition.

If you already have the skills to read and develop a profit and loss statement, feel free to jump to Part III. But would a quick refresher really be so bad?

The Basics Of Finance

Betty is the product manager for an office furniture company, and Frank is her CFO. Betty's been watching a trend where workers opt to stand all day in their offices instead of sitting in chairs because of the perceived health benefits. Betty goes to Frank's office. "I'm developing a proposal for a new line of executive standup desks. What do you want to see in the financials?" Frank says, "I don't need the weeds of a DCF model with IRR and NPV wrapped around it. Give me real data with 3-5 years of operating and cash flow projections."

Frank's response is typical of most finance executives and CFOs. Consider these points:

- They want specific information in a way they need to use it.

- They will use terms that they assume everyone else knows.

- They want to calculate key financial metrics, such as ROI, using their own desired methods.

- They are experts at developing financial statements and spreadsheets so they need the data.

A savvy innovator is very clear on the meaning and application of financial terms. At minimum, it's important to understand the most common tool used to evaluate the financial potential of a new opportunity—the pro forma profit and loss statement.

Pro Forma Profit And Loss Statements

A *Profit and Loss* (P&L) statement is the general scorecard used to determine the financial status for any business unit, product line, or product. A *pro forma* P&L statement is the primary tool used to evaluate new opportunities. 'Pro forma' is simply an estimation of the investment and cash flows in advance of having actual results. Once a pro forma P&L statement is developed for each year of the forecasting period, you have the basis for calculating a return on investment (ROI).

The format of a pro forma P&L statement is very straightforward. Looking at the sample in Figure 9.1, notice the five key elements:

Revenue. How much money will your company bring in during each period?

Cost of Goods Sold. The cost for making and delivering your product or service.

Operating Expenses. The range of research and development (R&D), sales, marketing, and other business expenses to execute on your new opportunity.

Income and Margins. The quantity of profit and their ratios to revenue at different levels in the P&L statement.

Return on Investment. The potential financial reward of an investment considering the value of money changes over time.

	Year 1	Year 1	Year 3	Year 4	Year 5
Revenue					
ASP $	23.00 $	23.00 $	23.00 $	23.00 $	23.00
Units	0	200,000	250,000	300,000	350,000
Total Revenue $	- $	4,600,000 $	5,750,000 $	6,900,000 $	8,050,000
Cost of Goods Sold					
Variable $	10.75 $	10.50 $	10.40 $	10.25 $	10.25
Fixed/unit $	2.75 $	2.50 $	2.50 $	2.50 $	2.50
COGS/unit $	13.50 $	13.00 $	12.90 $	12.75 $	12.75
Total COGS $	- $	2,600,000 $	3,225,000 $	3,825,000 $	4,462,500
Gross Income $	- $	2,000,000 $	2,525,000 $	3,075,000 $	3,587,500
Gross Margin/unit $	9.50 $	10.00 $	10.10 $	10.25 $	10.25
Gross Margin (%)	n/a	43%	44%	45%	45%
Operating Expenses					
R&D(9%) $	2,000,000 $	317,400 $	396,750 $	476,100 $	555,450
S&M (8%) $	500,000 $	368,000 $	460,000 $	552,000 $	644,000
G&A (6%)	$	276,000 $	345,000 $	414,000 $	483,000
Total Opex $	2,500,000 $	961,400 $	1,201,750 $	1,442,100 $	1,682,450
Operating Income $	(2,500,000) $	1,038,600 $	1,323,250 $	1,632,900 $	1,905,050
Op.Income/unit	n/a $	5.19 $	5.29 $	5.44 $	5.44
Operating Margin	n/a	23%	23%	24%	24%

Return on Investment	
Payback - Units	250,000
Payback Period	~15 months
NPV (14%)	$1,455,574
IRR	40%

Figure 9.1: A pro forma profit and loss statement

Let's take a look at each section and some common mistakes innovators make. Also look at the key questions decision-makers usually ask and how to answer them calmly and intelligently.

Revenue

Revenue is the amount of money your company will receive from the sales of your product or service. Revenue is NOT the same as cash. It can only be considered revenue from an accounting perspective once certain conditions are met, such as the customer has actually accepted delivery of your product, or your service has actually been completed.

Related Terms:

Net Revenue vs. Gross Revenue. Revenue has different meanings depending on circumstances. *Gross Revenue* includes all of the proceeds collected by the company. *Net Revenue* indicates a reduction of sales commissions, dealer discounts, or other direct sales expenses. Always specify what you mean by revenue. Using the wrong revenue will have a dramatic impact on later profit calculations.

Sales cycle/Accounts receivable cycle. Selling a product or service is not the same as getting paid for the sale. Understanding *cash flow cycles* means understanding when revenue is actually recognized by the company or when a customer will actually pay you based on their potential payment terms. Accounting for sales cycles may or may not be necessary for your financial model, but it's important to know if your executives need this.

Sell in/Sell through. For products that go through a distribution channel, your revenue forecasts should be based on a *sell in*— the quantity of product being sold into the channel. However, to forecast sales revenue over time, you must also consider a *sell through*—the quantity of products being sold through the channel to end customers.

Business model. The overall approach showing how your concept will lead to profit is called a *business model*. It might range from selling a product through distribution, selling direct to consumers, or giving an application away then making money on advertising, upgrades, etc.

Average Selling Price (ASP). Often a product or service is sold to different types of customers at different prices, like when large customers receive discounts. The ASP accounts for these types of customers or the range of customer prices on any revenue forecast.

Purchasing customer price vs. end-customer price. The *purchasing customer* is the entity that actually pays you. This may be a channel partner, such as a distributor, retailer, or value-added reseller (VAR). The *end-customer pricing* is the price given to the *actual* end-user of the product or service. These may be different customers, each paying different prices. You must be clear on which customer you are describing in your revenue model.

Cost. Beware that some people use *cost* and *price* interchangeably. Price is what customers pay when purchasing a product. Cost is what the company pays to create and deliver the product or service.

Common Innovator Mistakes:

Not reflecting actual net revenue on the revenue forecast. Many innovators make the mistake of creating P&L statements where sales commissions, volume discounts, and other significant items are ignored. I once watched an innovator present an opportunity for a new type of air quality monitor and used the end-customer price in his revenue forecast when he was planning to sell to a distributor. The price was off by at least 20%. He quickly realized his mistake, but it was too late to recover. Mistakes like these will ripple down through the P&L and create a wildly inaccurate estimate of the financial picture.

Not showing the relationship between volume and price. Volume will always highly correlate to price. Most innovators understand this, but can't describe how they are related. No one expects your first opportunity proposal to show a statistical understanding of price sensitivity, but you should be able to discuss how changes in price might impact demand or what price points must be achieved to drive certain levels of revenue.

Underestimating the timing to recognize sales revenue. Many P&L statements incorrectly show sales revenue as soon as a product is launched with no consideration of the time it takes to close customers and actually get paid. Business-to-business products, for example, may have a sales cycle that take months or years to go from an initial customer meeting to getting the first check. For example, selling into the education market can take up to two years waiting for customers to obtain budget approvals.

Having an unclear business model. At some point you'll have to ask the question, "How will we actually get paid for the concept?" This might be obvious, but the answer shouldn't be taken for granted. If your concept is similar to current products, then the business model will likely be the same. But for different concepts, you'll need to clarify this for executives. For example, a company that has traditionally sold hardware through distributors may not understand how software-as-a-service or advertising-based business models work. Be sure to explain these alternative models.

Cheat sheet to help you answer questions related to revenue:

Executive asks...	Executive thinks...	Non-savvy innovator thinks...	Savvy innovator responds...
"What makes you think these revenue forecasts are reasonable?"	(Do you even have a clue?)	(Can't you see how cool this is?)	"I've reviewed this approach with (executive) and given the existing data and current understanding of the customer value, the forecast seems reasonable."
"What assumptions have you built into your forecasts?"	(I assume you don't have a clue.)	(I assume you will never believe anything I say.)	"We used these assumptions for growth rate, relative value, adoption rate, and price estimates. We realize these are preliminary."

Cheat sheet (cont.)

"Why are you using this customer take rate?"	(Have you even talked with one customer?)	(What's a customer take rate?)	"Based on my initial discussions with customers and key factors in adoption, this take rate seems reasonable. We'll learn more in the next stage."
"What makes you think you can hit these volumes at this price?"	(Maybe some customers might pay for this…but really?)	(Geez already! How can I possible see the future?)	"I know price and volume are highly related. Based on the value customers placed on this concept and current market alternatives."
"What research did you use to get your numbers?"	(I need data!)	(How can we do research on something that doesn't exist?)	"We'll do a survey in later stages, but based on our customer interviews and competitor work so far."
"Isn't this just BS?"	(I don't believe you.)	(What do you want from me??)	"At this stage, of course these are rough estimates. What is clear is that we can be early in solving a real problem that customers will pay for."

Cost of Goods Sold

Cost of Goods Sold (COGS) includes the costs and related expenses to build and deliver each unit of product or service. COGS are different than the research and development expenses used to design and prepare the product for manufacturing, which are included in operating expenses. COGS can be difficult to estimate when you don't know much about actual features and functions. However, when discussing profit margins you must try to estimate what COGS would be at product launch and how they'll change over time and at various volumes.

Related Terms:

Cost of Sales. Similar to COGS, but used more often for services or other non-product businesses. This includes all direct and variable costs as well as allocations of costs that are indirectly related to developing the service.

Bill of Materials (BOM). A BOM includes the cost of the parts and materials used in a product and also items such as the product warranty, manuals, and packaging.

Direct Costs. These are costs directly associated with producing the product or delivering the service such as the BOM, manufacturing expenses and direct labor.

Indirect Costs. These are costs indirectly related to producing the product or delivering the service such as overhead for electricity, office space, or purchasing agents as well as the costs for operations managers. These costs are often given an allocation to specific products or services based on your company's selected accounting principles.

Fixed Costs. Fixed costs are those costs that don't change based on volume such as the cost of creating molds, tooling, or setting up a manufacturing line.

Variable Costs. Variable costs are those costs that directly relate to each unit produced or delivered. Variable costs mostly consist of parts and labor.

Labor. The cost of human labor to make or deliver a product or service.

Overhead/Burden. The portion of a company's general expenses that are allocated to a product or service. These often include shared utilities, rent, general management, supporting functions, and other office expenses.

Depreciation. A cost over time for the expense of large purchases, like equipment, that eventually get used up and become obsolete. For example, if you purchase a mold to create a new mobile phone case, and that mold cost $150,000 and lasts three years, you would depreciate the expense over three years and apply $50,000 to either COGS or expenses for each year.

Common Innovator Mistakes:

Ignoring components of the whole product. A product has many elements beyond the physical product or direct service, such as warranty expenses, packaging, or direct customer support expenses. However, these are often left out of COGS.

Underestimating the real production costs. A big problem when communicating COGS is ignoring many of the costs that are associated with the real world of producing a product. Innovators often underestimate or forget the cost of shared manufacturing facilities, such as the overhead costs of related personnel or the shipping costs associated with delivering a product to distribution.

Calculating COGS incorrectly. Estimating COGS is a cost accounting effort that factors hundreds of direct and indirect costs. Often you just don't have visibility into everything that goes into COGS, such as burden rates and allocations for various expenses. You must work with finance colleagues to get acceptable COGS data.

Not communicating the relationship between cost, volume, and investment expenses. The initial COGS of a new concept might be very high, but volume can force those costs to come down significantly. Likewise, you might find ways to reduce the product cost by investing more in production or R&D. Note these relationships. They're critical elements of your financial conversations.

Cheat sheet to help you answer questions related to COGS:

Executive asks...	Executive thinks...	Non-savvy innovator thinks...	Savvy innovator responds...
"Are there suppliers for all of these parts?"	(Do we need to invent a lot of new things or be at the mercy of one supplier?)	(Isn't it too early to tell suppliers about this?)	"We don't have specific suppliers, but we've identified several candidates we'll evaluate in upcoming phases."
How can we get these costs down?	(There is no way we'll make money at these costs.)	(Of course they'll come down once we sell a million of these.)	"COGS is high at launch, but will come down when we hit these volumes. We can also cost reduce the design once we see market success."
"Have you considered all of the overhead costs?"	(Do you really understand all the expenses related to delivering this thing?)	(I just gave you all the BOM and manufacturing costs. What else do you want?)	"Yes. I reviewed these with our cost accounting group."
"Do your COGS include everything?"	(Do you know how to develop COGS?)	(Of course they do. I included the BOM and assembly costs.)	"Yes. I reviewed the whole product using a similar product as reference."
"Who will be building this?"	(Our company is not building this thing...are we?)	(We can build anything... what does she mean?)	"We need to review this as we learn more about the design and manufacturing needs."
"What does the manufacturing team say?"	(Please tell me someone who actually understands this has reviewed it.)	(What? It's too early to tell manufacturing about this. It's just an idea."	"We had a preliminary discussion and they helped with the numbers."

Operating Expenses

Every cost of doing business that doesn't come out of revenue or COGS shows up on the P&L as an expense. The expense line is challenging. Getting help from a knowledgeable colleague is critical. Technical people usually have little idea how much sales and marketing will be required to drive sales. And likewise, sales and marketing people usually don't know what it takes to develop a product.

For existing products and services, operating expenses are usually allocated as a percentage of revenue applied across all products. For example, one company allocates 6% of revenue to every product for line item 'General and Administrative Expenses' to cover office space, professional salaries, and utilities. For new opportunities, you need to determine if these allocated expenses are acceptable or you may need to develop specific expenses related to the new concept.

Related Terms:

Sales and Marketing Expenses. All of the expenses related to selling and marketing that are not already accounted for in calculating net revenue. These usually include advertising, press relations, websites, and brochures, as well as sales and marketing salaries, commissions, and bonuses.

Research and Development (R&D) Expenses. The operating and capital expenses necessary to develop, prototype and test a concept from initial conception to launch, including all employees and capital purchases expenses.

General and Administrative. The various expenses incurred in running a business that range from electricity to executive salaries. These usually stay relatively flat over time and are allocated across all products and services.

Capital Expenses. Capital expenses are typically for purchasing equipment that can be depreciated over their useful life. The calculated depreciation becomes part of operating expenses. If capital purchases are required for the opportunity, the finance team can help determine the best way to include these in a P&L.

Fully-burdened Employees. Employees cost more than their salaries, which simply means other employee-related expenses must be included such as benefits, stock options, computers, and office space.

Common Innovator Mistakes:

Underestimating research and development expenses. Most executives will assume you're underestimating R&D expenses overall, especially for new concepts where your team has little experience. Innovators tend to assume development will go faster with fewer problems than is normally the case. They also often underestimate the time and resources necessary for testing and fixing problems in follow-up development efforts.

Underestimating sales and marketing expenses. An ongoing product usually gets allocated a percentage of revenue for marketing, but a new product or new market will require significantly more resources to generate demand in early periods. This should be accounted for with additional sales and marketing expenses.

Underestimating or excluding manufacturing or other start-up expenses. Any new product or service will have a range of set up expenses ranging from training customer service agents to writing product manuals. Often major items are left out of investment estimates.

Inappropriate accounting treatment of expenses. You should be careful when including capital expenses or making assumptions about what should be included as either an expense or as part of COGS. The finance team can help.

Not including all employee expenses. Hiring additional people to research, develop, manage, and/or market the concept costs more than just their salaries. Be sure to include all of the associated expenses in your estimates.

Cheat sheet to help you answer questions related to operating expenses:

Executive asks...	Executive thinks...	Non-savvy innovator thinks...	Savvy innovator responds...
"Do you really believe these R&D estimates?"	(You guys always underestimate both time and resources needed.)	(Why can't you just trust me?)	"Based on what it will take to create a competitive product, we worked with engineering to estimate."
"Why have you used those S&M expense estimates?"	(Do you have a clue as to what it will take to create demand?)	(This is the budget we use for other products...what's wrong?)	"Based on what we think will be required to find customers and create demand, this is our best estimate."
"Has finance reviewed your expenses?"	(Please tell me you had someone that understands finance review this.)	(How can some bean counter who doesn't understand this idea help?)	"Yes"
"Have you included all the company overhead expenses?"	(These look low to me.)	(What?)	"Yes"
"How have you accounted for capital expenses?"	(Do you know what you're doing?)	(What?)	"I worked with finance and they helped me include them in either COGS or expenses as appropriate."
"Isn't this just BS?"	(I don't believe you.)	(What do you want from me??)	"At this stage, of course these are rough estimates, what is clear is that we can be early in solving a real problem that customers will pay for."

Profit and Margins

A funding conversation focuses a lot on margins. Executives believe, rightly so, that if you're not able to sell the product or service at some reasonable profit then there's no point in reviewing the opportunity further. This is a challenge for innovators since so much information is unknown. Profit depends on every other estimate of price, COGS, and expenses.

You must be able to speak intelligently about profit, margins, and income regardless of how detailed your information is at this point. Often executives want to see a very specific number, like a gross margin greater than 40%. Learn these financial hot buttons to defend how your concept will achieve these numbers or why you believe the targeted numbers don't apply to your idea. Figure 9.2 summarizes the relationship between revenue, COGS, operating expenses, and profit levels.

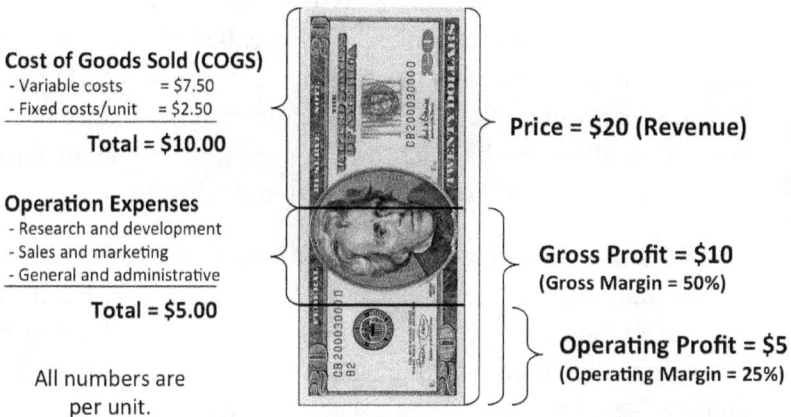

Cost of Goods Sold (COGS)
- Variable costs = $7.50
- Fixed costs/unit = $2.50

Total = $10.00

Operation Expenses
- Research and development
- Sales and marketing
- General and administrative

Total = $5.00

All numbers are per unit.

Price = $20 (Revenue)

Gross Profit = $10
(Gross Margin = 50%)

Operating Profit = $5
(Operating Margin = 25%)

Figure 9.2: Income and margin overview

Related Terms:

Cash. The term cash has a very specific meaning for accountants. Sometimes executives mean cash and sometimes they mean income, sales revenue, or something else. In any case, when you hear or use the term cash, be clear on what is meant.

Gross Income/Gross Margin. Subtracting COGS from net revenue leaves gross income. Gross income expressed as a percentage of revenue (calculated by gross income/net revenue) is gross margin. These are both important financial metrics for every executive. You must be crystal clear as to these meanings and calculations.

Operating Income/Operating Margin. Subtracting expenses from gross income leaves operating income. Operating margin is expressed as a percentage and is calculated by operating income/net revenue. Operating margin is important, but for new opportunities, it's not usually as important as gross margin since many corporate expenses are unrelated to a specific opportunity.

Net Income/Net Margin. Subtracting taxes and other expenses from accounting or financing activities leaves net income. The net income of a P&L statement is also called the *bottom line.* Net margin is expressed as a percentage and is calculated by net income/net revenue. Net margin is important to the finance team, but not usually something innovators need to worry about since they have little control over accounting or tax issues.

Contribution Margin. Similar to operating margin is contribution margin. This metric is calculated by removing the fixed costs from the COGS. It shows how a product or service is contributing to the expenses of the whole company or other products and services. I personally don't see this used much in evaluating new opportunities, but it is important in some companies.

Earnings Before Income Taxes, Depreciation and Amortization (EBITDA). This has become a popular metric to evaluate startups because some people think it represents a better view of a company's actual cash flows. EBITDA is a good example of something important to know if your executives think it's important to know.

Common Innovator Mistakes:

Not understanding margin goals. Knowing margin targets and how your opportunity meets these goals is critical to success. For example, if a company averages 40% gross margin across all current products and your idea is estimated to hit 30% gross margin, then it's weighing down all other products from a financial perspective.

Using inaccurate profit terms. Many people use the terms margins, income, and profit interchangeably. But since profit is so important to the discussion, it's critical to be accurate when you use this and related terms. Saying "we expect profit of 25%" is less accurate than "we expect operating margin of 25%."

Estimating unrealistic margins. It's easy to calculate large margins by manipulating the underlying variables of revenue, COGS, and expenses. But margins must be based on reality. If the numbers look too good to be true, they probably are. Use other companies and products as comparisons to make sure margin estimates are of the right magnitude.

Cheat sheet to help you answer questions related to profit and margins:

Executive asks...	Executive thinks...	Non-savvy innovator thinks...	Savvy innovator responds...
"How can you get your gross margin to over 50%?"	(This doesn't come close to meeting our margin goals.)	(Isn't 35% enough money for you? Talk about greed.)	"I realize this is lower than our typical margins. Here are some possible options considering price and costs."
"What makes you think you'll hit that operating margin?"	(There is no way you'll hit that operating margin.)	(Why can't you just trust me?)	"I've reviewed these numbers with the finance team. What area gives you the most concern?"

Cheat sheet (cont.)

"How does this compare to other products in this category?"	(How can we possibly compete at the margins you're showing?)	(How am I supposed to know that?)	"I estimate that competitors have a similar cost, but lower operating expenses because…"
"Why is your gross margin going up in later years? Shouldn't they go down?"	(Doesn't she know margins will always erode over time?)	(Don't you want them to go up? What's the problem?)	"Gross margin is low initially due to our high costs. I agree prices will come down but cost will come down faster."
"Can't you just raise the price to get the margins up?"	(Raise your price to get the margins up.)	(OK.)	"We could raise prices, but this may seriously affect volume. Let me take a look at that scenario further."
"Isn't this just BS?"	(I don't believe you.)	(What do you want from me??)	"At this stage, of course these are rough estimates. What is clear is that we can be early in solving a real problem that customers will pay for."

Return On Investment (ROI)

Ultimately you'll be asked, "What return on investment can we expect from this opportunity?" The irony is that there is no standard ROI calculation, so you must understand what ROI means to your decision-makers. It could mean a specific calculation, such as the result of a Net Present Value (NPV) or Internal Rate of Return (IRR) calculation, or perhaps some other specific metric used internally to calculate an ROI. In many companies, the finance team doesn't even want innovators to develop ROI calculations because the finance team will do this. That's ok, let them. However, you need to understand the results and how they'll be perceived by decision-makers.

Related Terms:

Discounted Cash Flows (DCF). Basically, money today is worth more than money tomorrow because of interest rates and inflation. Therefore, cash flows in the future should be reduced by some established discount rate. The discount rate is usually calculated based on the interest rate of borrowing money from a combination of banks and investors.

Hurdle Rate. Every investment should achieve some minimum rate of return. This threshold is called a hurdle rate. Any return below this rate is considered a bad investment and won't support your company's needs. You must know what your company's hurdle rate is for your opportunity.

Cash Flow Breakeven. The point when the cash flow from investing in the opportunity exactly matches the cash flow from operating income. After this period the opportunity begins to contribute profit to the company.

Net Present Value. The value of an investment in today's dollars, taking into account the initial investment and all of the future discounted cash flows using a specified discount rate. NPV is the most common calculation used for new opportunities. By definition, an NPV of greater than zero is a good investment, yet other opportunities may have a higher NPV. NPV provides a magnitude of what the opportunity might be worth to the company, but doesn't provide an idea of the actual rate of ROI.

Internal Rate of Return (IRR). This is the financial return that an investment will produce in terms of a percentage. Generally you can view it as the interest rate you'd get on the investment. For instance, if I put $2 million in the bank (the investment) and received $1 million in cash over the next three years (and nothing in the fourth year), the IRR on this investment would be a 23%. (You may need to use a spreadsheet to validate this.)

Payback period and number of units sold. Payback period is the time it takes to pay off the original investment. If I invest $5 million in year one and have positive cash flows of $2.5 million in each of the following years, then the payback period will be two years after the initial investment period. If the gross margin of each unit is $10, then the number of units before payback is 250,000 units. Note this number does not include expenses. You'll have to determine how your company calculates payback if it's an important metric.

Common Innovator Mistakes:

Using the wrong ROI formula. Calculating NPV and IRR of an investment is easy using a spreadsheet, but be certain you are using the correct ROI formulas that your company needs you to use. This includes using the cash flow line your finance team expects you to use such as either operating income or net income.

Using the wrong discount rate. NPV and other ROI calculations require that you use a designated discount rate. Some companies like to vary the discount rate based on the type of opportunity. For example, with low risk opportunities you might use a discount rate of 12%, but a higher risk concept might require a 20% or higher discount rate. You must learn how your executives want you to account for risk in financial calculations.

Getting overly excited about a huge NPV or IRR. You may be able to calculate an IRR percentage in the thousands. Looks like a home run! But, a huge IRR is rarely believable and discredits the rest of your presentation. Focus on the underlying projections and let any ROI speak for itself.

Cheat sheet to help you answer questions related to ROI:

Executive asks...	Executive thinks...	Non-savvy innovator thinks...	Savvy innovator responds...
How did you calculate your ROI?	(Do you have a clue on how we calculate ROI here?)	(I used the Excel spreadsheet formulas.. what's the problem?)	"I worked with finance and they calculated ROI based on my income estimates using..."
"What discount rate did you use?"	(Do you understand the risk of this opportunity?)	(Umm...is this a trick question?)	"I used 13% based on what finance told me was reasonable for this type of opportunity."
"When will we get our investment paid back?"	(I don't want to see anything longer than two years.)	(I just showed her the ROI is huge. What does she want?)	"The payback period is about 18-24 months. We'll have a better estimate when we learn more about the R&D expenses."
"How can we accelerate the payback period?"	(This is way too risky if you take this long to make money.)	(Just give me a lot more money.)	"There are a couple of options to accelerate the time to market or increase early sales."
"Which income line is your ROI based on?"	(Do you have a clue?)	(What do you mean? I used the income line that shows profit.)	"I used the operating income line based on the finance team's direction."
"Isn't this just BS?"	(I don't believe you.)	(What do you want from me??)	"At this stage, of course these are rough estimates. What is clear is that we can be early in solving a real problem that customers will pay for."

SAVVY IDEA: NEVER PRESENT YOUR OWN FINANCIAL MODEL

It's Thursday afternoon and Ted is fifteen minutes into his funding presentation. The meeting has been going well and his audience seems very excited about the opportunity. He's now ready for the big reveal of his financials. He hits the next page on his presentation and the eyes in the room squint to see the cells of his carefully manicured P&L spreadsheet blazing in front of the room. Ted proudly starts sharing the numbers, "Based on the market data and our work with sales and customers, we believe we can hit these numbers." Clara, the CFO looks over the spreadsheet and sees that something is not right. Pointing to the spreadsheet, she declares, "That cell doesn't look right. You're showing a 32% gross margin in year three, but it looks more like 25%. Clara is right. There is something wrong with Ted's calculations. At this point Ted is doomed. Nothing on his spreadsheet now is believable. He might as well stop the meeting and regroup after he has checked his numbers.

You might be very handy with a spreadsheet, but guaranteed your financial team is even more skilled. There are simply too many chances for calculation errors with margins, ROI, allocations, etc. Even if your calculations are correct, your finance team may compute them differently. My advice is to always work with a respected financial person to develop the actual spreadsheets. Not only do you have a significantly greater chance of gaining acceptance of the numbers, if there is a problem with the spreadsheet you have an ally who you can ask, "What do you think? Should we go back and look at that closer?" I promise, they have the skills to quickly recover from any financial oddities and can explain the problem more accurately than you.

CHAPTER 9 REVIEW—A 30-MINUTE CORPORATE INNOVATOR MBA

It's impossible to explain every financial term or variable you'll encounter. But a savvy innovator is familiar enough with the language to effectively work with financial teams and hold intelligent, executive-level discussions. Try not to let the intimidation of complex financial statements worry you. There are no right answers at this point of opportunity development. A calm, confident response like "I don't know at this point, but here is my plan," will go further than an exasperated walk through a complex spreadsheet that no one trusts. If you're not fluent in financial language, find a respected finance person. He or she can help you understand and develop the pro forma P&L statement that your company needs in order to see the financial potential of your BIG idea.

Five Savvy Success Strategies

As you hone your skills, practice these five strategies to deepen your understanding of key financial concepts and terms.

1. **Take a finance class.** Taking a class (or two) to build your financial skills can save you a lot of time and aggravation. It also looks good on your resume. A program that focuses on reading and creating P&L statements and understanding basic cost accounting is a good place to start.

2. **Make friends with finance colleagues.** Not only can finance colleagues provide much of the information you need to develop accurate financial statements, they can also be powerful allies when trying to sell ideas.

3. **Be clear on internal finance terms.** While financial terms often have more standard definitions than marketing terms, every company uses them differently. To be considered an insider within your company, always be clear on the most important financial terms used by your company.

4. **Don't try to recover from bad calculations.** Inevitably someone will find a flaw in your spreadsheets, but attempting to locate and rectify it in front of an audience is a recipe for disaster. When this happens, calmly state, "I'll go back and take a fresh look." And then do.

5. **Be certain that financials match your story.** It's easy to make a spreadsheet say anything you want. But stating you will, "offer a premium product in the market" and then showing average prices with low margins in a spreadsheet damages your credibility. Always make sure the elements of your story are consistent and match every line of your financials.

BOB AND NANCY (CONTINUED)
Bob scrambles to pull it all together.

Bob has only a few days left to develop and deliver a proposal that will impress Nancy. He's tempted to just send her the 100 slides he has so far to see if he's on the right track, but (smartly) decides against it. He knows that Nancy needs strong numbers, so he focuses next on cleaning up his financials.

He sets up a meeting with Bridget, a senior financial analyst, and then emails Nancy to clarify a few questions.

"Hi Nancy, I've been working on the HM-3000 proposal and I think it's going well. Keith has been great and led me to Jenny for help with customer interviews and to Chris for help with forecasts. I'm preparing for our meeting next Friday and have a few last questions. I know you're busy, but any guidance is appreciated.

- How much data do you expect behind the revenue forecasts?

- I interviewed five customers so far, is this enough for the first proposal?

- How long do you want the proposal to be?

- How detailed do you want to see the development forecast?

Thanks much, Bob"

Nancy responds that evening.

"Bob, Glad you're getting help and it sounds like you're on the right track. I need to cancel our meeting, but we have the monthly operations review next Tuesday. Have my assistant slot you for 20 minutes at the end. As for the proposal, just share the concept, customer value, market forecasts, and financial summary. See you then. Nancy"

When Bob reads the email he thinks, 'Well that wasn't much help. I've only got four days left, and now I'll be presenting in front of a whole group of executives. What am I going to do?'

The next morning Bob meets Bridget for their meeting.

Bridget: *Hi Bob. It's nice to meet you. How are you doing?*

Bob: *Well, to be honest, I'm freaking out at the moment. I need to give a presentation for a new opportunity at an operations review meeting on Tuesday. I was getting help from Chris, but he said my financials were a mess and I should see you.*

Bridget: (Geesh. Why do these guys always wait so long to see me?) *Let's see what you have.*

Bob pulls out his presentation, but Bridget stops him.

Bridget: *Let me see the actual spreadsheet.*

Bob opens the spreadsheet on his laptop and Bridget's eyes dart around the cells. She picks a column and scrolls down stopping briefly at every formula.

Bridget: *Well, I can see what you're trying to do, but that's not a format we use. Where did you get these costs and expense estimates?*

Bob: *I just took a best guess. This is just a concept right now so I figured estimates were fine.*

Bridget: *Estimates are fine, but you can't leave out major costs or expenses. Are you planning to produce this at MedCo?*

Bob: *I assume we will. Is that important?*

Bridget: *Well, you haven't included a lot of our internal costs and expenses. Assuming that your revenue is close to right, your margins are way too high.*

Bridget continued to explain the structure of a P&L and pulled up a P&L for a current product that included line after line of costs and expenses with names that Bob didn't understand. He tenses up. (Holy crap, I'll never get this done.)

Bridget: (Pulling up some other spreadsheets.) *This is a report generated from our accounting system. This is a pro forma P&L for another opportunity we looked at last year. It should simplify things for you. Let me send you this. Just follow the format and send it back to me when you're done and I'll take another look. OK?*

Bob: (Feeling the tension release from his body.) *That would be awesome. I can't thank you enough.*

Bridget: *Oh...one more thing. You're showing you need an investment of $3.2 million for the first year. I hope you're not asking for all of that in your proposal.* (The phone rings.) *I need to take this. Good luck.*

Back in his own office, Bob ponders her last statement and thinks, '$3.2 million is the lowest I could estimate for R&D. If I ask for less we won't have a product. Hmm...what am I missing?'

Then another thought occurs to him, 'What happened to that last opportunity Bridget mentioned and the person who presented it? I must know!' His hands start shaking at the thought of falling flat on his face. At this point he's not as concerned about his idea as he is about losing his job.

WHAT'S GOING ON?

Bob is at a crossroads. He's facing a meeting in four days that can make or break his reputation as an innovator. The good news is that he has a lot of information now to present. The bad news is that this will be his first presentation to the executive team and his presentation is a mess.

Can Bob pull the presentation together in time to save his idea, and also save face?

THE GRAND FINALE
Bob faces the executive inquisition.

It's Tuesday morning. Bob is sitting outside the boardroom waiting for someone to let him know when it's time to present his big idea. He can feel his throat getting dry. The door opens and he's motioned in.

Bob says nothing as he heads to one end of the table and sees Nancy on the other side. Each of the eleven executives look up from their phones and day planners to stare at Bob as if they were waiting for a magician to start performing.

Nancy breaks the silence, "Hi Bob. Let me introduce you to the group while you're setting up. Bob came to me over a month ago with an interesting idea for a new remote heart-monitoring product. I asked Bob to present the idea today. Afterwards, we'll discuss it and get back to him with next steps. Ok. Bob, the floor is yours."

As Nancy is talking, Bob starts to calm down. He's impressed with how she's controlling the room. She even mentions the name of the idea! Bob begins his presentation.

Bob: *The concept I've been working on is to add wireless communication and cloud services capabilities to our current heart monitors and to start a new line of remote heart-monitoring devices. These will target the hospital market segment and will solve a pressing problem for the doctors we interviewed.*

He goes to the first slide that shows a doctor walking in an airport looking at a heart signal on his mobile phone. A patient is shown in a nursing home lying comfortably in her bed.

Bob: *I'm looking for $150,000 to develop a rapid prototype by modifying our current device. I also want to create a mobile application that displays a heart being monitored we can* *use for testing, and I also need to purchase some additional industry reports.*

Bob: (Clicking to slide three that shows a simple graph.) *Trends show that healthcare providers want to save money by keeping patients at home; either in private homes or minimum care facilities. This is a rapidly growing market segment. Remote monitoring is only $140 million today, but projected to be over $2 billion in 5 years. Based on the doctor feedback we received, the biggest benefit would be the flexibility to monitor patients from anywhere they happen to be—even the golf course.* (The audience chuckles at that.)

Bob: *We interviewed doctors who are the primary influencers in purchases. They told us the key monitoring capabilities they want. We can reasonably modify our equipment and integrate with a cloud service to meet their needs.*

The first question comes from Chris. He already knew the answer, but wanted to throw Bob a softball to show his support. "Isn't our competitor, Remote Diagnostics, already doing this?" Without missing a beat, Bob clicks to the next slide that shows an industry landscape in a 2x2 matrix.

Bob: *Yes, Remote Diagnostics does provide remote monitoring. But their sensors still use wires to connect the patient to the monitoring equipment. I'm still researching their roadmap, but I believe we can be first with wireless sensors. This would be a significant advantage since doctors told us that patients often lose or disrupt the sensors when they roll around in bed. Doctors don't trust them unless a technician has just placed them on the patient. What's the benefit then of remote monitoring? Wireless sensors will build that trust.* (The crowd nods along with Chris.)

Bob continues fielding questions while trying to stay on target. Keith's suggestion to shorten the presentation to half the time allotted turned out to be great advice.

Bob explained the technology in less than a minute and showed a slide of the system architecture that highlighted three key areas for investigation and his plan to research them. The sensors, the wireless protocol, and the battery technology were critical. His final slide showed three parallel paths for the technology, customer research, and go-to-market planning. Each path included a milestone in 30 days and then additional milestones at 60 and 90 days. The $150,000 would cover the resources he needed to finish the prototype in 90 days.

Nancy: (Looking at Keith who's been silent.) *What do you think Keith? I understand you've been working with Bob on this.*

Keith: *It makes sense to me. I reviewed it with Bob yesterday. There are obviously a lot of details to work out and the cost and revenue projections are anyone's guess right now, but I don't see any reason not to pursue it further.*

Nancy: (Looking at Ted, the head of sales.) *Ted? What's your opinion?*

Ted: (Subtly shaking his head.) *I don't have a firm opinion yet. I agree this fills a market need, but I told Bob early on that Wi-Fi doesn't seem like the best path. I've seen products like this that have failed. I won't trust the technology until I see it working in the field.*

Nancy: *Ok. Bob, make sure testing is part of your plan. I also want to know why other products have failed. I know you're asking for $150,000 to do a proof of concept and further research. Thanks for breaking this apart from the whole project. Here's what I'd like to do...I'll approve $75,000, and I want to see the additional industry data and your progress on the prototype in 30 days. We'll review your progress to decide about the other $75,000. Does this give you enough to develop the next phase of a research plan? Five doctors are* *nice but we'll need more. How much time can you devote to the project?*

Bob: *I can devote about 30% of my time based on other projects. We can get more feedback from other doctors as well as several large nursing homes that our hospital customers say we should work with. I don't know how much we can do in 30 days.*

Nancy: (Quickly scanning the faces for further comments.) *Well, do your best. Unless there are any objections...you have the approval to go ahead. We'll see you in 30 days.*

Bob is elated, but knows not to say anymore. He walks out with a big smile and wonders how he's ever going to thank Keith, Chris, Bridget and Jenny?

WHAT HAPPENED?

Bob spent his last four days giving himself the best chance for success. Working with Keith and Chris as his coaches, he developed a concise, compelling opportunity proposal that was backed by as much data as he could find at this preliminary stage. He also got help from experts to address the major challenges of risk, customer value, and revenue forecasts. Their help was invaluable, but so was their support and credibility in the meeting. Bob also pre-sold the proposal to his managers, who could have been a big hurdle if Bob had not involved them in advance of the operations meeting. Also by limiting the resource requirements through a clearly defined MRA, Bob lowered the risk for everyone.

While not all requests for funding go this smoothly, he carried out all the right steps to build a credible case and survive the executive inquisition. Bob, along with his BIG idea, is well on his way to becoming a savvy innovator.

PART II CONCLUSION

The goal of Part II in *The Savvy Corporate Innovator* was to develop the building blocks that savvy innovators use to sell their ideas in the murky world of corporate innovation. We explored the language of innovation, how to address the personal and business risk of a new opportunity, and how to answer two primary questions:

- Do customers care?
- Should the company care?

The story of Bob and Nancy also provided a running dialogue to share how one innovator gained support for his exciting new medical product. By working through the key elements of his idea and seeking expert help along the way, Bob was able to successfully thrive during the executive inquisition.

Part III, titled The 30-Day Action Plan, pulls all of the previous concepts and information into a step-by-step action plan to take an idea from a mere spark to a compelling opportunity proposal that meets the needs of funding executives. We'll also follow another case study and see how Mark fares with his MossBeGone robotic roof-cleaning idea.

PART III

THE 30-DAY ACTION PLAN

Steps For Developing And Selling A Concise Opportunity Proposal

Chapter 10: Prepare For Success
Clarify the Opportunity and Plan Your Work

Chapter 11: Interview Potential Customers
Identify and Interview People for Your Customer Needs Pyramid

Chapter 12: Estimate The Revenue Potential
Creating Market and Revenue Forecasts that Persuade Executives

Chapter 13: Develop A Preliminary Go-To-Market Plan
Address Product Development, Marketing, Sales, and Other Operations Tactics

Chapter 14: Determine The Value To The Company
Put the Financial Story Together, Including Pro Forma Financials and ROI Estimates

Chapter 15: Make The Proposal Concise, Compelling And Complete.
Fill In Missing Gaps to Create a Succinct Proposal that Includes Your MRA

Chapter 16: Sell The Proposal
Develop the Specific Tactics for Selling Your Idea to Company Executives

---

INTRODUCTION TO PART III

If you assume every statement will be met with,
'Prove it to me!', then you'll do fine.

By now you're well aware that you must be able to articulate and sell ideas with an opportunity proposal that excites decision-makers. In Parts I and II, we've been exploring how to examine your strengths and weaknesses, categorize your ideas, evaluate your company's opportunity environments, and understand your executives' needs. Savvy innovators take all these factors into account when preparing for the executive inquisition. Now you'll be able to put that knowledge into action. Part III provides specific tactics, activities, and tools in a 30-Day Action Plan to develop a concise, compelling proposal.

As you work through each chapter, you will address both sides of the funding challenge: 1) Developing a compelling proposal that defines and validates your BIG idea and 2) Taking steps to gain executive support that helps achieve your initial funding goals.

The 30-Day Action Plan will keep you on track as you navigate your way to success. Each chapter provides a step-by-step guide along with specific goals, activities, and the types of data you'll need to gather. You'll also follow Mark, the innovator I introduced earlier, who wants to create a moss-removing robot for his company. Mark's own experience will help you envision how each step of the 30-Day Action Plan can be implemented.

After 30 days, you'll have an opportunity proposal and a selling plan for moving your idea to the next stage of development. Remember, don't strive for a perfect proposal—work to meet the needs of funding executives.

Chapter 10

PREPARE FOR SUCCESS

Clarify The Opportunity And Plan Your Work

Your selling process begins the moment you share your idea with someone else.

Savvy Steps To Success

Step 1: Identify the Company's Proposal Format

Step 2: Conduct a Rapid Market Scan

Step 3: Prepare to Unveil the Idea

Step 4: Interview Executives

Step 5: Consider the Prototype

Step 6: Plan Your Work

Introduction

Once you have a BIG idea, you must then plan for success. You may be tempted to start telling others about your idea, but if you try to sell it too fast, you may start the corporate antibodies in motion and have a harder time gaining support later. While your idea may only be a spark of inspiration at this point, before sharing it, get clarity and consider everything you know and don't know about it. You don't want to tell anyone about your idea until you're certain they'll understand it without forming a negative opinion.

This begins with gaining insight into the company's objectives and executive needs. Once you have this insight, you'll have the basis for developing a solid 30-Day Action Plan for building your opportunity proposal. Let's get started!

Step 1: Identify The Company's Proposal Format

This may sound obvious, but a good first step is knowing exactly what is expected when you present your opportunity proposal. Most companies have a template available for submitting new opportunities. It may be called a "business case" or "idea proposal," and the sections may vary from company to company, but they're all designed to clarify the concept, show its market potential, and explain its connection to the business. The bad news is these templates can be frustrating because they often request information that may not be relevant at this early stage of development.

If you're required you to use a standard template, that's OK—use it—if that's what decision-makers want to see. Sometimes it's the executive making a decision who created the template in the first place. However, don't just simply fill in the blanks. A savvy innovator knows that a standard template should be treated seriously, but also never assumes that a well-written proposal will lead to funding success. The standard template might not provide the order and exact information you need when presenting your opportunity. You may have to customize it to create a compelling story; not just walk through the boiler-plate sections of a template.

Elements of an opportunity proposal

Executive Summary: A bit of steak and a lot of sizzle to intrigue the audience to read further.

The Concept: A short description of your idea that focuses on the customer along with a visual story to get excitement started.

Target Market: Clarification of how you have defined and segmented the market. This is where credibility is built or destroyed and the questions start flying.

The Value Proposition: The part where you must convince others your idea solves a real and pressing problem, fills a serious need or has some other clear value for customers.

The Market Potential and Forecast: The financial potential of your idea, which shows you understand the market as well as executive needs.

The Execution Plan: The preliminary how, who, what, and when of your plan.

Product: The preliminary product description and development plan.

Channels (Place): The preliminary sales and distribution plan.

Pricing: Preliminary price estimates.

Promotion: The preliminary marketing communications plan.

Pro Forma Financials: The financial numbers to justify the investment that must be supported by relevant data.

Risk Considerations: A clear summary of internal and external risks and your plans to address them.

The Opportunity Score Card: A score that shows how your idea ranks using company investment criteria.

The MRA and Next Steps: The minimum resources you need from the company to take your opportunity to the next level and how you will use them.

Back Up Materials: All that other information that isn't absolutely necessary to tell your story, but may be important to risk-averse decision-makers.

Step 2: Conduct A Rapid Market Scan

Selling ideas is very much a "chicken and egg" problem. You don't want to go deep into analysis without first getting feedback, but you need some information to speak intelligently about it.

Before unveiling your idea, get a cursory understanding of the current market situation. Later, we'll take a more rigorous look at the market to develop forecasts, understand customers, and analyze competitors, but for now, start by asking these questions:

- What is the general market size for the type of concept you're thinking about?
- What are the most obvious competitors?
- Are there any recent announcements or events that relate directly to your concept?

Your goal is to briefly scan the market so when you do have a conversation about your idea, you have at least some awareness of what is going on.

Step 3: Prepare To Unveil The Idea

The selling process starts now. To gain support for your idea, decision-makers must see the value for both customers and the company and see this value through murky, risk-averse eyes. As discussed earlier, savvy innovators set up short meetings with executives to talk about an idea, get them onboard, and understand their needs. However…before meeting with any executive or even your manager, be prepared! Be able to succinctly describe your idea and know exactly what you want to achieve during those early discussions. Next are two exercises to help prepare you for talking about your idea. Taking time to develop these two items will not only prepare you, but will also show that you're very serious about pursuing your opportunity.

One Page Opportunity Summary (OPOS). A summary of the concept to help clarify the elements of your idea and develop an initial communication tool. An OPOS forces you to create a hypothesis that you will then use to test, refine, and validate your thinking. It provides something tangible for both early stage discussions as well as an executive summary for comparing your idea with other ideas.

There is no perfect one-page format. At this point, your information may be merely educated guesses, but the exercise alone will help flesh out details and see any gaps in your thinking. When I'm working on new ideas, often just creating an OPOS gets me excited to keep moving forward, or allows me to kill the idea before wasting time. What's important is that the concept is clear in your head to prepare sharing the idea.

20-second opportunity pitch. A short, intriguing summary of the opportunity to plant seeds and get initial meetings set up with executives. This is similar to the popular "elevator pitch" you're probably familiar with. Your goal is to intrigue your audience with a clear, compelling and potentially valuable opportunity— not just an idea! The audience must be able to quickly relate to the concept and see how it could benefit the company. An opportunity pitch might sound like this:

"I've been thinking about [Concept]." What if we were able to [key benefit] for [target customer]? Currently they [customer problem]. Based on [data], I think this could be [opportunity]."

Now fill in the blanks...say you have a pet peeve about dry erase markers for white boards. They always start OK when they're new but quickly lose their ability to write clean, dark lines. You've never found a good one that keeps writing well for more than 10 minutes without losing its pep. Now say you're the product manager for a dry erase product company and just had an idea to develop a new premium dry erase marker. Here's your 20-second opportunity pitch:

"I've been thinking about a new premium dry erase marker. What if we were able to develop a marker that maintained its ability to write over long periods for professionals that rely heavily on whiteboards for communication? Based on my quick research, there are at least 1.2 million educational instructors and consultants in the US alone that I think are likely target customers. I believe this could be a $20 million opportunity for us. I'd like to research this further."

Your pitch is obviously more compelling the more specific you can be, even with just one piece of compelling data. But if you don't have any data or specific customers in mind, that's OK for now. Use your 20-second pitch to ask for permission to investigate further and plant some seeds to harvest later.

Step 4: Interview Executives

A sure way to fail is to have an idea, spend three months working out the details, prepare a big presentation, then blindly go into an executive inquisition. Your chance for success? Almost zero. The odds that you've addressed the key issues, gathered the right data, and gained political support are astronomical. This is why earlier I stressed the importance of talking with executives to understand their needs.

Interviewing decision-makers early in the proposal process is the step that most innovators miss, but it is critical to success. By taking action now to sit down with funding executives, you'll start to understand their specific needs and get to know them on a personal level. You'll also gain insight into the company's innovation goals and financial situation. There's also a practical reason for meeting executives early on—scheduling meetings can take days or weeks. You may be well on your way to a finished opportunity proposal before your first meeting with an executive. That's OK. When you do meet them, you'll be even more prepared for an intelligent conversation.

How many executives to interview

I recommend interviewing two or three executives at this stage. You're looking for preliminary information to help focus your efforts. This will save you days (or months) gathering and analyzing data. Ideally you first meet the specific decision-maker that will make the final decision and ask, "Who else do you think I should meet with?" but it's never that easy. Your goal is to identify the most influential executives you can approach. Some are more accessible and open to ideas than others. Who you approach also depends on other factors. For example, with a new market opportunity, you'll want to meet with sales and market executives. For a technology opportunity, try to speak with the head of development and CTO.

And please...don't forget your direct manager is this process! He or she should be the first person you meet with to discuss the best approach for moving your idea forward.

Prepare interview questions in advance

Before sitting down with your manager or any executive, prepare a short list of questions. You may only have 15-20 minutes, so use the time wisely. You're trying to understand their attitude toward risk, any specific data they need to see, and clues for how to remove perceived risk related to your idea. Below are some suggestions:

- What are the two or three criteria most important to you to approve funding?
- Tell me about the last several opportunities you've reviewed. What was good or bad about them?
- What was your first reaction to the idea I just shared?
- How do you think this opportunity fits with the company today?
- Is there anything specific you'd like me to address as I develop the proposal?
- How would something like this fit with your goals?
- Are you interested in being a mentor as I develop this opportunity?

Throughout this process, consider the functional background of each executive. Ask questions related to their world to uncover their personal criteria. Find out what they worry about most. Get as specific as possible. The key to understanding is in your follow-up questions. My favorites are:

- Why is that important to you?
- Please tell me more about that (i.e. can you be more specific)?
- What would success look like?

Get clear on each executive's hot buttons

If you think you already know each executive's needs, you can skip interviews, but do so at your own risk. Most companies have stated criteria for evaluating opportunities such as revenue potential, profit, and technical risk, but many executives also have personal 'hot buttons' you need to know about. For example, I once worked with one very experienced COO of a consumer electronics company. His background was not in electronics, but in marketing fast food munchies like potato chips. He had seen a lot of great successes and failures, and his experience told him that opportunities live and die by conducting research using test markets. This meant actually putting new products in the market to see if they would sell. Only then would he believe the risk was acceptable to invest in full-blown commercialization. This executive's 'hot button' was customer testing, and anyone wanting support from him would need to have a clear plan for customer testing in a proposal.

For first time innovators

Obviously the closer you are to executives in the corporate hierarchy, the easier it is to get their time, but it's often as simple as asking the right way. If you sound confident and keep their needs in mind, you'll usually succeed. Think back to your 20-second opportunity pitch: "Hi Linda. I'm putting together a proposal for... (insert 20-second opportunity pitch here). I'd like to set up 20 minutes to get your thoughts before I get too far on the proposal. Would that be ok?" I've rarely seen this not work. Most executives want to hear about new opportunities. They also want to stay fresh on trends, technology and ideas. And they're always looking for sharp people that know how to execute.

Help! My manager is a bottleneck

Innovators often complain that their direct manager acts as a harsh filter for their ideas. When you try to present an idea, your manager says, "No thanks!" or promises to think about it and then nothing happens. If this is your situation, the one thing you cannot do (except under extreme circumstances…like changing jobs) is to go around your manager and present the idea to his or her boss. Doing so creates a very dangerous political situation for you. Instead, try one of these options:

- Uncovering the real reasons your manager is saying no, then working with them to get a yes by following the steps in this book.

- Acknowledging the resistance then asking for permission to seek counsel from a colleague, such as, "I really understand your concerns about this idea. I'd like to meet with Jim (the COO perhaps) and see what he thinks. Are you OK if I set up a meeting with the three of us to discuss?"

- Trying to use the internal processes, like working with your manager to get the idea as an agenda item on an operations meeting or asking if there's a forum where you can get feedback.

Whatever you decide, always keep your manager informed of your progress and activities. If you're having success, make him look good in the process. If you're failing, don't blame her for not supporting you. Your goodwill will pay off in future efforts.

Step 5: Consider The Prototype

It's helpful to mockup a simple prototype of your idea to show executives how it will be valuable for a customer. But there are many levels and variations of prototypes, so knowing the right level to develop for a 30-Day Action Plan is critical to getting support for your idea.

"Hey Jim, can you draw that idea you had for the cooking application on this piece of paper for me?" "No problem Jesse. Do you want me to do a wireframe? Or, I could do a quick demo in Flash for you." Jim and Jesse just discussed three levels of prototypes. Each level would have varying amounts of detail, achieve different goals, and require increasing amounts of time, resources, and attention.

One challenge is simply the word 'prototype'. It's viewed by many as a near-complete and functioning product. Depending on the complexity of your idea, it could take years (and many millions) to develop a working prototype. But for an initial proposal, you don't need anything close to functioning. Think of your prototype as a communication tool and not proof that you can actually create a functional product. You will have to prove the technology at some point, but right now you aren't trying to validate your concept.

One innovator I worked with was able to sell an idea for a new health application by creating a short animated video before having any idea how to build it. Your prototype can be as simple as a drawing, a foam mockup, or a couple of graphics. Steve Jobs is famous for creating a foam mockup of the Apple iPod and carrying it around his pocket to see how it would feel. Prototypes obviously get more sophisticated as the product evolves, but for the 30-day Action Plan, keep your energy on the prototype at a minimum.

A FEW PROTOTYPE IDEAS FOR THIS FIRST STAGE OF DEVELOPMENT

Consider these potential prototypes for various types of ideas:

For a software or web application product: Use simple sketches with minimal graphics and text to explain the two or three primary benefits of the application and how it might be used.

For a physical product: Make a foam or cardboard mockup. With 3D printing, creating a physical prototype is easier and cheaper than ever.

For a component of another product: If the idea is an electronic or mechanical component of another product, using a preliminary marketing sheet that clearly highlights important benefits and attributes is reasonable.

For a service: Draft a preliminary brochure that shows a visual of a customer using the service.

An example of an early prototype is shown in Figure 10.1. This is one panel of a storyboard that was used to show how consumers might use a new music application in their car to enhance (or possibly distract from) their driving experience. (This testing was done before the wide spread use of smart phones.)

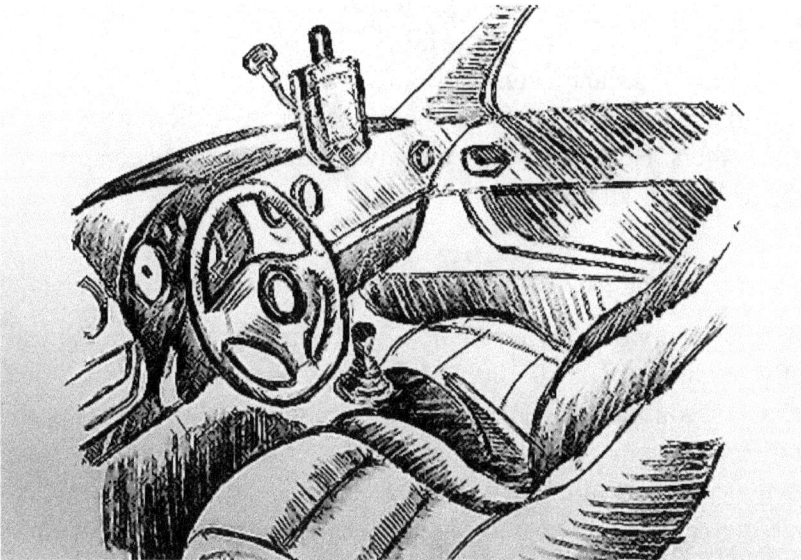

Figure 10.1: Simple graphics make great prototypes

Prototypes can be too good

One last thought. A simple, compelling drawing or mockup of your idea enhances your discussions with executives. However, if your prototype looks too good, then people might think you're showing the finished product. They may want to discuss the specific colors, button locations and other details. Then you have to explain, "But this is just a prototype!" Keeping your prototype more conceptual right now will actually lead to better and more objective feedback.

Step 6: Plan Your Work

Now you've got a clearer picture of your concept and the market situation. It's time to plan your work and schedule your time. This will be your systematic path forward that defines your specific goals and activities to keep you on track. Don't let your idea linger and become one of the 212 sticky notes of "things-to-do" covering up your computer, demanding your attention. And don't let your attention be drawn towards product features, technical aspects, and implementation details of your concept, while giving only a passing thought to market analysis and actual customers. I've certainly been guilty of this. It's tempting to get your concept " just right" before sharing it with others. However, a savvy innovator takes a balanced approach that considers all other factors.

Factor in time for your project

A focused opportunity evaluation for the first stage of funding usually requires about 80-100 actual work-hours—some more, some less. As you gain experience, you'll be able to plan activities and estimate time more precisely. It's always difficult to determine how long it will take to find data, set up customer interviews and schedule meetings with executives. These activities can, and should, be accomplished in as close to 30 days as possible. Hot opportunities can't wait!

Opportunity: A summary of the idea		Start date:	Day 1
		End date:	Day 30
		Status	Blocked! Behind On track

Key Objective: Prepare for Success	Key Activities - Clarify the opportunity - Interview executives	Est. Time ~10 Hours	Due Date Day 1	Status On track
Key Objective: Interview customers	Key Activities - Prepare to interview customers - Set up interviews	Est. Time ~20 Hours	Due Date Day 5	Status On track
Key Objective: Estimate revenue potential	Key Activities - Gather relevant data - Develop financial model	Est. Time ~24 Hours	Due Date Day 10	Status On track
Key Objective: Develop execution tactics	Key Activities - Develop product and R&D plan - Develop S&M tactics	Est. Time ~14 Hours	Due Date Day 15	Status On track
Key Objective: Develop pro forma financials	Key Activities - Estimate costs and expenses - Develop financial P&L	Est. Time ~10 Hours	Due Date Day 20	Status On track
Key Objective: Complete the proposal	Key Activities - Create simple graphics - Edit and refine	Est. Time ~10 Hours	Due Date Day 25	Status On track
Key Objective: Sell the proposal	Key Activities - Set up review meetings - Prepare for final presentation	Est. Time ~10 Hours	Due Date Day 30	Status On track
Total Estimated Time ~100 Hours				

Figure 10.2: Key elements of a 30-day action plan

Success requires the right balance of activities

Exactly how you'll execute each activity over the next 30 days may not be clear yet. That's OK. We'll discuss each step thoroughly in upcoming chapters. For now, simply identify the major tasks and list them in their priority using a basic timeline. Figure 10.2 shows a sample list for each element of a 30-day Action Plan along with time estimates for each set of activities.

Ultimately you're trying to answer the question, "Why this idea over all others?" With that in mind, don't worry about spending time doing exhaustive analysis or gathering numerous details. Keep your work simple and focus on developing a concise opportunity proposal in the time allotted, with as much relevant data as possible. Try to line up activities in a logical order and with the right amount of attention to hit your primary objective—a compelling pitch that meets the needs of funding executives.

Innovators aren't exactly known for following project plans, so it's up to you to schedule a plan that works best for your style and environment. You'll never follow a plan exactly, because as soon as you develop one, the situation changes. As Dwight Eisenhower said, "Plans are nothing. Planning is everything."

CHAPTER 10 REVIEW—PREPARE FOR SUCCESS

After coming up with a BIG idea, innovators often get really excited and want to jump right in and start developing. But for *real* success (that is, getting funding for the idea), it's important to take a step back and prepare. You need a clearly articulated concept, an understanding of what executives want and need, and a 30-Day Action Plan. Once you take the necessary steps to achieve these things, you're well on your way to funding success.

Savvy Success Timetable

Times will vary by project and executives' needs, but a savvy innovator should use a balanced approach to address every aspect of an opportunity. As a guideline, here are the estimated hours you should spend following these steps of your 30-Day Action Plan to prepare for success:

Step 1: Identify the company's proposal format	1 hour
Step 2: Conduct a rapid market scan	4 hours
Step 3: Prepare to unveil the idea	2 hours
Step 4: Interview executives	2 hours
Step 5: Consider the prototype	2 hours
Step 6: Plan your work	1 hour
Total:	**12 hours**

MossBeGone
Mark's 30-Day Action Plan

Remember Mark who we met at the start of this book? Mark is a savvy innovator and an engineering manager for RoboCo, a robotics company based in Seattle. Mark has a BIG idea to propose a self-cleaning, moss-removal robot for entering the consumer market. We'll follow Mark as he prepares for the executive inquisition and carries out his 30-Day Action Plan for his MossBeGone opportunity.

Objective 1: Prepare For Success

It's Monday morning, day one of Mark's efforts to get support for his moss-eating robot. He's excited about his MossBeGone concept and believes the product he envisions has great potential for RoboCo. Mark already has a full-time job managing his engineering department, so he'll need to squeeze in his innovation activities to develop a compelling opportunity proposal. He begins following these steps:

Step 1: Identify the company's proposal format

Mark goes to the company's intranet and downloads the template for an opportunity proposal. It looks close to the one described earlier in this chapter. There are no surprises, except for several line items that ask for specific customer information such as: 'Describe the customer requesting the product.' and 'Estimate the size of the expected order.' These make sense since RoboCo produces high-end manufacturing robots customized for large business customers. The template could help him understand his company's needs, but Mark knows he'll have to take a different approach and create a unique proposal for a consumer robot.

Step 2: Conduct a rapid market scan

Next Mark focuses on the market. He researches the Web for publicly available information and finds the following:

- Based on US Census data there are 132 million homes in the United States, of which 75% are single-family homes.

- The market for roof maintenance and supplies is about $1.2 billion.
- The consumer robotics market is growing at 14% annually, whereas the industrial robotic market is flat.
- There were no obvious competitors selling an autonomous roof-cleaning device. However there was one mechanical product that cleaned roofs, but it had to be controlled manually by an operator on the roof.

Mark also finds two research reports that might provide more data:

- "Consumer Robotics Market: 2014-2019"
- "Roofing Construction and Supply Market Statistics"

There is a lot of advice about roof-cleaning problems, but very little data that directly relate to the size of the market for roof-cleaning. He'll come back to that issue later.

Step 3: Prepare to unveil the idea

Based on his new knowledge of the concept, value for customers and the market, Mark develops his One-Page Opportunity Summary as shown in Figure 10.3. He has to make educated guesses at this point, but thinking about each section gives him more insight on how to develop a project plan to complete his proposal.

He then scripts his 20-second opportunity pitch:

"I've been thinking about a new product for expanding into the consumer robotic market that is growing at 14% a year. What if we could make a small autonomous machine, similar to robotic vacuum cleaners, that could maneuver on a roof to remove moss? Based on my early research, moss removal is a $1.2 billion problem in the United States alone that takes a lot of time for homeowners and uses potentially hazardous chemicals. We could use our robotic technology and moss-detection sensors to create a better and safer solution. I'd like to research this further and come back with an opportunity proposal in 30 days."

MossBeGone
One-page Opportunity Summary

The Ask:
Prototyping resources - TBD
Market research resources - TBD

The Situation

The Big Idea:
An autonomous roof cleaning device using robotic capabilities to remove moss and other roof contaminants

Target Customers:
Homeowners in high-moss regions

Market Size:
~100 Million single family US homes
$1.2B in roof servicing and supplies
Target segment is TBD

Supporting Trends:
Robotics technology
Autonomous devices
Advanced sensors

Competitive Landscape:
Developers of other consumer-oriented autonomous devices
Current roof cleaning services and devices

What created this opportunity?
Personal observation of the problem.
Discussions with other homeowners.

The Strategy

Value Proposition:
Automatically remove moss from roofs: Save significant time; save roof maintenance; save money

The Revenue Model:
Sell wholesale to distribution network; large home supply and hardware stores; online direct sales

Market Attack Elements:
Heavy use of video testimonials and demonstrations; viral marketing plan TBD
Target price <$300

The Numbers
TBD with 30 Day Action Plan

	20XX	20XX	20XX	20XX	20XX
Units	TBD	TBD	TBD	TBD	TBD
Rev.	TBD	TBD	TBD	TBD	TBD
Exp.	TBD	TBD	TBD	TBD	TBD
OI	TBD	TBD	TBD	TBD	TBD

The Tactics

Solution Summary:
Small autonomous robot with high roof traction; sensors for moss and edges; moss killing agents; mobile app for status updates

Development Plan:
Leverage current robotic technology; identify sensor partners TBD
1st product: 18months, Est. $5m?
Cost target: $<150?

Customer validation plan:
Work with home improvement partner to validate and test

Key Initial Milestones:
- Proof of concept &validate value: 6 mo.
- ID/close channel partners: 12 months

The Scorecard (Scale 1-5)

Criteria	Score
Strategic Fit	4
Revenue potential	4
Time to profit	?
Technology Risk	3
Commercial Risk	2
Total	?

Figure 10.3: MossBeGone one-page opportunity summary

Step 4: Interview executives

Now Mark wants to approach his executives to better understand their needs. He prepares this short list of questions to ask his manager and decision-makers:

- Have we looked at other areas of robotics outside of manufacturing? What happened?
- What are your initial concerns about the consumer robotics market?
- What are the criteria most important to you to approve funding?
- What is your first reaction to developing a roof-cleaning device?
- Is there anything specific you'd like me to address as I develop my proposal?
- I'd like to review the proposal with you as I get closer to completion. Is this OK?

Mark already has a pretty good working relationship with many of the executives. However, presenting a really new opportunity to them will be a new experience. So to better prepare himself, he meets with his manager, Linda, the VP of R&D, and shares his idea. Although earlier she'd been skeptical and said she didn't think Grant, the CEO, would go for it, she decides to support him investigating further. She tells him it's OK to plant some seeds with the executives and work to understand their needs. Linda was actually looking forward to getting this information herself. With her green light, he sets up two meetings using his 20-second pitch and goes right to the top to request a meeting with Grant. Since Mark knew RoboCo was very financially driven, he also sets up a meeting with Tony, the CFO.

During those meetings the next week, Mark learns the following:

- The CEO and CFO like the idea, but they had initial objections like, "How could it go across different levels of roofs?" and "Wouldn't it need to carry a huge reservoir of chemicals?"

- Any opportunity would need to have potential of at least \$25 million in revenue within four years.
- They want to see positive cash flow by year three.
- The CFO needs to see a serious market analysis to believe in the concept.

This all makes sense to Mark, but he's not sure yet how he'll convince his executive team that MossBeGone can meet these criteria.

Step 5: Consider the prototype

Prototyping proves difficult for Mark. Not because he can't do it, but because he wants to do more of it. He has the vision so clearly in his mind, it would be easy to put two of his team members on it and have a working prototype in six months. The resistance to do this is incredible. Instead, he decides to spend his personal time evaluating the robot technically, and work with one of his team members to create a graphic depiction of a moss-eating robot scurrying around a moss-infested roof.

Step 6: Plan your work

Mark then works on planning his tasks and scheduling his time for the next 30 days. While he's very comfortable developing plans for his technical projects, developing a plan for MossBeGone turns out to be more challenging. He has to estimate how long it will take to gather data, find customers and conduct analysis. He is new to all of this. But he's up to the challenge. He develops a simple spreadsheet to document a 30-day Action Plan as shown in Figure 10.4.

Mark's next challenge is to get a clearer picture of the value that MossBeGone has for potential customers.

(To be continued…)

Project Lead	Mark
Opportunity Description	An autonomous roof cleaning device using robotic capabilities to remove moss and other possible roof contaminants
Project Success Metrics	1) Customer validation with 10 consumers and 5 retailers 2) Clarify and quantify the target market 3) Financial criteria: • Cash flow positive by year 3 • Revenue of >$25M by year 4

Project Activities	Due	Status
Project Kickoff	Day 1	
Set up customer interviews	Day 8	
Gather data and analyze the market	Day 15	
Identify and analyze competition	Day 17	
Estimate market potential and revenue	Day 18	
Develop preliminary execution plan	Day 20	
Develop pro forma financials	Day 22	
Develop preliminary proposal	Day 24	
Preview proposal with 3 executives	Day 27	
Develop final proposal and presentation	Day 29	
Opportunity review	Day 30	

Day 1 — Start!
Day 8 — Customer Plan Complete
Day 17 — Analysis Complete
Day 24 — Proposal 90% Complete
Day 30 — Opportunity Review

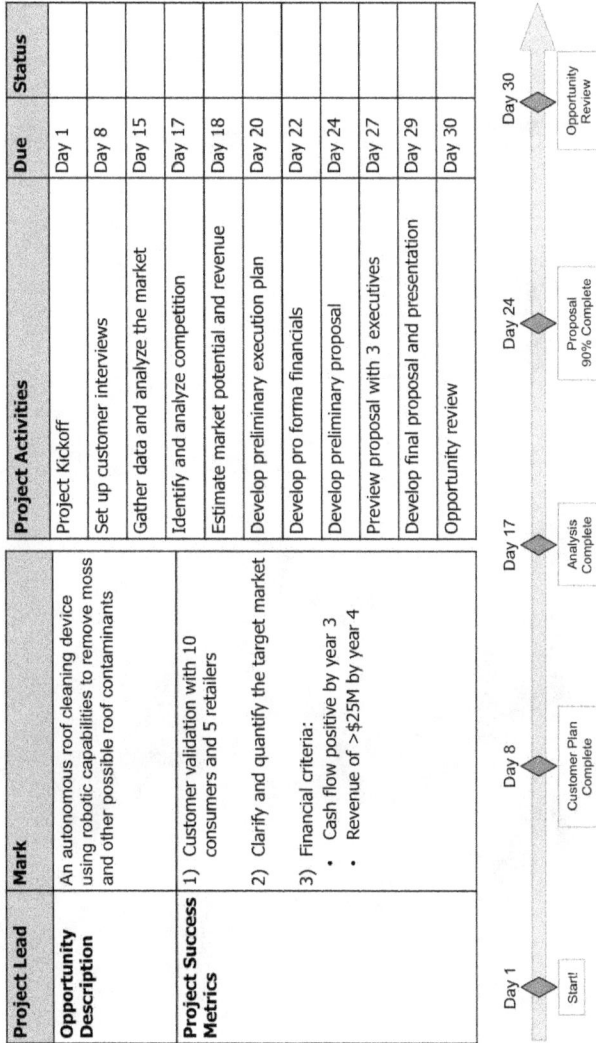

Figure 10.4: MossBeGone 30-Day Action Plan

Chapter 11

INTERVIEW POTENTIAL CUSTOMERS

Identify And Interview People For Your Customer Needs Pyramid

---

Once you start interacting directly with customers, you'll wonder why you've never done it before.

Savvy Steps To Success

Step 1: Develop a Customer Persona

Step 2: Find Suitable Customers to Interview

Step 3: Develop a Needs-First Interview Guide

Step 4: Conduct Probing Interviews

Step 5: Process Results and Develop a Customer Needs Pyramid

Introduction

When you present your BIG idea at the executive inquisition, you'll be up against opposing opinions from executives with a great deal of authority and seniority. Many innovators assume their idea, along with all of its cool features and technology will sell itself. But, without compelling and relevant data to show that customers will value and adopt your concept, a funding decision will only come down to your opinion versus theirs. And the opinion of those with more power will always prevail. To get funding, you must present compelling data obtained directly from potential customers. As discussed in Chapter 7, the best technique for getting this data is interviewing a small number of targeted customers.

Let's get started.

Step 1: Develop A Customer Persona

Once you've identified your target market, your next step is to clarify the characteristics of the customer you want to find and interview— the target customer. One way to do this is with a customer persona. A customer persona is not a person, but a profile that represents a typical customer in your target market. For now, it's ok to have one persona that represents your entire target market. As your opportunity evolves, you'll need to develop different customer personas that define the range of customer types including users, purchasers, and channel members.

The persona is based off a draft you develop from your initial hypothesis of your target customer's characteristics. After you conduct interviews, you'll refine your draft and add details to finalize the persona that you'll present with your opportunity proposal. To develop a draft persona, visualize the most likely end-customer who has the problem your idea addresses, then summarize that customer's situation and why he or she needs something better. Once you have a firm description for your customer profile, create and add a visual to your opportunity proposal.

Once you have a draft, your next step is to connect with customers who match your hypotheses.

Step 2: Find Suitable Customers To Interview

Your goal at this stage is to explore customer needs rather than try to thoroughly validate your idea. However, it's very difficult for people to articulate what they need. The best approach is to identify those customers who can explain their needs and provide feedback on ideas that are currently just concepts. An example is a category called 'lead users'. The concept of a lead user was developed by Eric von Hippel of MIT in 1986. He defines lead users as advanced users who seek out and can benefit from innovation. They often create new solutions to solve their own problems. These customers are aware of market conditions, have thought more deeply about their problems, and can articulate their needs better than the average customer. Lead users are perfect candidates to gain an early read on the value of your new idea. With consumer products, lead users are often called 'early adopters'. They're the first ones who research and buy the latest and greatest products.

On a cautionary note, many innovators make the mistake of getting feedback from channel partners first, such as retailers or distributors, as opposed to end-customers. Channel partners can provide good insight, particularly with how to bring a product to market, but they're not the ultimate end-customer. Interviewing end-customers is the best place to start to get the fastest and most accurate insight into customer needs.

Get help finding potential customers

Getting access to suitable end-customers requires a little leg work, persistence and ingenuity. Start by looking within your personal network of industry colleagues, friends, and family. As a corporate innovator, you have the advantage of tremendous access to potential customers through your coworkers. Remember, you don't want to use colleagues *as* your potential customers, but sales, business development and marketing coworkers can all help you with the process of *finding* customers. This is especially important if your idea is for a new business-to-business opportunity where finding suitable people to interview inside a large company can be challenging.

If you have direct access to customers, it's tempting to call a couple of current customers, share your idea, and see if it resonates with them. But this tactic has a downside. Rarely will you get a true understanding of those customers' needs. They may react favorably to your idea just because of your relationship—and this won't be genuine feedback. To obtain relevant data that executives find credible, go beyond talking with only those customers you already know. You need to be objective when getting customer insight or risk developing a proposal that appears heavily biased.

Sales people usually have the largest customer database and may be your best source. However, sales people are often concerned about sharing what they might perceive as a "half-baked" idea because their reputation is on the line. They're responsible for building great customer relationships and hitting revenue targets, so it's natural they'll be protective of those relationships. To put sales people at ease, assure them that you are making NO promises to their customers; you're just asking some questions to explore their needs.

Tradeshows, sales conferences, distributor locations, and other venues are also places to find potential customers. One innovator I worked with, Hans, developed a new card game for kids that included animal statistics so kids could "fight" each other's animals to win or trade cards. Kids could learn while playing. Using prototypes of animal cards, Hans conducted two types of customer insight activities: he stood outside a game store (with the owner's permission) and played the game with kids to see their reaction as well as their parents' response. He also went to local schools and shared the game at recess—watching and learning as kids played. Based on this early insight, Hans refined the game, and it ended up being a big hit. He later sold it to a major game company for a tidy sum!

When you seek help from others, ask to be introduced to people who match your customer persona. You don't need to explain your entire concept, but do share the problem you're researching. For example, if you're researching a new type of security camera, you might ask, "I'd like to talk with a couple of people who have valuable possessions and want to ensure they are safe."

How many customers do you need?

I recommend you complete five customer interviews. Five solid interviews will exceed most decision-makers' expectations because they rarely get any solid customer data with initial proposals. To get five interviews within 30 days, you should request an interview with at least ten customers that match your profile since about half will usually fall through. Finding ten potential customers for a new concept shouldn't be difficult. If it is, this may be a clue that finding customers to buy your concept will also be difficult.

Step 3: Develop A Needs-First Interview Guide

It's essential you prepare a set of questions before interviewing customers. This "interview guide" will help keep your interviews consistent and on track. I've personally interviewed thousands of potential customers, and while there's no perfect interview guide, the good ones follow a common framework and help prevent a common problem known as *confirmation bias*.

Confirmation bias is our tendency to hear what we want to hear in order to confirm a belief we already hold. This makes it easy to reject any data that goes against our ideas. Innovators often try to sell a concept instead of really listening to potential customers, peers, and executives. One huge mistake is showing a potential customer a prototype and asking, "Is this something you would buy?" If the customer doesn't respond with a clear NO, then any other answer (and potentially relevant feedback) is interpreted as a Yes!—without taking into account what the customer is saying. To overcome confirmation bias, a savvy innovator starts an interview by learning about the customer's situation, problems and needs before sharing any specific information about the idea. This is a 'needs-first interview' approach.

As you can read in the next examples, you'll get a great deal more information by leading your customer interviews to understand their needs first. The idea is to let them explain their situation to you without first trying to sell them on your product. Once you've learned their issues, you can then prioritize their needs, determine if your concept fulfils their most important needs. And understand any objections to adopting the idea.

THE AUTOMOBILE INTERVIEW: TWO APPROACHES

The year is 1885. You've just invented something called an "automobile." There are two different ways you might initiate an interview with potential customers. The first method starts by focusing on the product:

You: "Thanks for coming in. I'd like to talk with you about automobiles today."

Customer: "What's an automobile?"

You: "It's like a horse and carriage, but you don't need a horse."

Customer: "Then how will my carriage go?"

You: "It has an engine that takes gas."

Customer: "What's an engine? Do mean the gas my horse always has?"

You: "Uumm, no, not exactly…say, let me show you a picture… look at this."

Customer: "Darn, that's an ugly horse. I don't want one of those! It looks complicated. I'm happy with my horse."

After this unsuccessful discussion, you make a note: "Customer doesn't get it. Find smarter customers."

The second, more fruitful conversation starts with understanding a customer's problems and needs first. This is a needs-first interview and goes something like this:

You: "Thanks for coming in. I'd like to talk to you about how you get back and forth to town with your family."

Customer: "Great!"

You: "Tell me about the last couple of times you came to town."

Customer: "Well…last Saturday, I hitched up and we came in for supplies."

You: "Great… tell me more."

Customer: "Let's see. I had to get a lot of supplies, so we used the four-horse team and the big carriage. It takes about four hours, so we got an early start because the kids get scared after dark. We also packed a lunch because I didn't want to have to stop once we got into town. It also rained like crazy that day. The wife got wet and was freezing and ended up catching a cold. When we got home that night, I had to take care of the kids. She was worthless. Now I'm thinking about getting a new top for my carriage. You know, one of those new ones where you can seal up the sides."

At this point, you can now introduce the concept of your automobile and have an intelligent discussion with the customer based on the needs he's just expressed. You can discuss the value of speed for getting to town, the value of more horsepower to carry heavy supplies, and the value of an enclosed cabin to keep out the weather. All things the customer might find valuable in an automobile.

Developing an interview guide—the 4x25 rule

Successful interview guides follow a 4x25 rule. Basically you develop a set of questions that share time equally among the following areas:

Explore the Problem: Use 25% of the time with questions designed to get to know the potential customers, understand their situation and explore problems that your idea addresses.

Explore Customer Needs: Use 25% of the time with questions designed to explore what is important to them when seeking solutions and making decisions.

Explore Customer Value: Use 25% of the time with questions designed to discuss your idea and determine if it fulfills important needs. This may include sharing a paper prototype or just explaining your idea.

Explore Issues and Objections: Use 25% of the time with questions designed to understand how customers might visualize adopting your idea and any issues that concern them. Allow them to talk about what they want to talk about. Often this is when they'll open up and tell you what they really think.

An interview guide is not meant to be a rigid set of questions, but a guide to keep you on track and ensure you cover important topics. Some of the top questions I like to ask are shown in each category below:

Category	Sample questions
Explore the Problem	Tell me about a time when you… How big a problem is… Walk me through how you approach… How do you feel about current solutions and providers today when…
Explore Customer Needs	What are the most important results you need to see when… What is important to you when looking for a way to… What frustrates you when you think about… Based on what you said is important, how would you prioritize…
Explore Customer Value	What if you could purchase a device that could… I'd like to see what you think about… How would you compare this idea to… From a scale of 1 to 10, how would you rate this for your situation… What would make it a 10?
Explore Issues and Objections	What major concerns would you have about something like this… What would you be willing to pay to… What would prevent you from adopting something like this… What do you really think about this idea?

This guide can be used for a short chat or a two-hour discussion. I like to keep interviews to around 60 minutes, but if it's a good conversation with a customer who really gets into it, let it go as long as needed.

Important!
Real insight comes from probing questions

The savviest innovators use probing questions to follow up on a customer's initial response. This technique is critical and should be used throughout an interview to go deeper into the meaning beneath a response. These are the same probing questions you would use during your executive interviews:

- Tell me more…
- Why did you say that?
- Why is that important to you?

Once you have developed a clear interview guide, you're ready to conduct a focused customer discussion that will generate maximum insight in the shortest time.

Step 4: Conduct Probing Interviews

I remember my first customer interviews. Even with an interview guide, there were many awkward pauses, and I found it hard to listen and probe because I was too busy preparing the next question in my mind. However, after about three or four attempts, I got the hang of it and actually started to enjoy them

Listen for needs, not features

One of the skills I had to learn was how to probe for needs when a customer wanted to discuss features. It's easy to get caught up discussing features and technology, but your best results come from focusing on customer needs. It's critical to stay on track.

When customers say they want or like a feature, they may really be signaling a need that requires a solution. It's essential to uncover what that need might be. Think back to Joan and the BananaEnhancer from Chapter 8. She might ask, "Tell me what you need to see in a device that extends the life of a banana." The customer says, "I need it to have a glass door." However, a glass door is a feature, not a need. Joan wants to understand why the customer needs a glass door. If she probes further, "Why is that important to you?" she might learn that the customer really needs an easy way to see how ripe the bananas are at any moment. So rather than using glass, Joan might be able to solve this need using a clear plastic door or no door at all.

Do face-to-face interviews when possible

Ideally, conduct your interviews face-to-face so you get the benefit of body language, intonation, gestures, and facial expressions. Phone interviews will suffice if time is of the essence and customers aren't in your local area. Interviewing customers does take some practice, but is well worth your investment to build this skill. For the best results, follow these tips:

- Practice on several people before your first real interview. Friends and colleagues are good guinea pigs.

- Allow room for silence to let both you and the customer gather thoughts.

- Record the conversation so you don't miss anything because it's very difficult to think, write, talk and listen at the same time. But always ask permission first!

- Bring a teammate if possible to take notes and discuss your findings.

- Document everything immediately after the interview. If you wait too long, you'll only remember what you want to hear.

Step 5: Process Results And Develop A Customer Needs Pyramid

Once you've conducted a series of interviews and have pages of notes, it's time to process the results. Always be aware of confirmation bias, otherwise you'll damage your credibility. It's easy to cherry pick the information that supports your idea and ignore everything else to conclude—"Customers love it!"

A better approach is to draft an objective summary that starts with the customer problems you heard and a clear explanation of their important needs. Organize those needs into priorities using the Customer Needs Pyramid (CNP) we discussed in Chapter 7. Once you have your CNP, you can assess whether your idea fulfills these needs by objectively answering the following questions:

- Does my concept fulfill the needs that matter most to customers?

- Do customers see real advantages in my idea when compared to alternatives?

- Can I overcome the issues and objections that customers have raised?

- Does this idea create enough value to warrant a first round of funding?

For now, you don't need definitive answers, but your interviews should provide enough data to form educated opinions. You'll continue to answer these questions and develop better answers in upcoming steps. Just keep in mind that your goal is not to statistically validate your idea at this point, but to confirm it's moving in the right direction. Show executives that customers value your idea, that there are no essential needs that can't be fulfilled, and that your concept is worthy of a first round of funding.

CHAPTER 11 REVIEW—INTERVIEW POTENTIAL CUSTOMERS

By interviewing five customers in the idea development process, you'll acquire powerful, important data for your initial proposal. You'll learn what needs customers have when they attempt to solve the problem your concept is trying to solve. Your preliminary research won't be perfect, but it will give you a much better understanding of how to be successful in the market as well how to gain the support of executives.

Savvy Success Timetable

Times will vary by project and executives' needs, but a savvy innovator should use a balanced approach to address every aspect of an opportunity. As a guideline, here are the estimated hours you should spend following these steps of your 30-Day Action Plan to interview potential customers:

Step 1: Develop a customer persona	1 hour
Step 2: Find suitable customers to interview	4 hours
Step 3: Develop a needs-first interview guide	2 hours
Step 4: Conduct probing interviews	10 hours
Step 5: Process results and develop a CNP	3 hours
Total:	**20 hours**

MossBeGone
Mark's 30-Day Action Plan (continued)

Objective 2: Interview Potential Customers

Mark is ready to start the process of interviewing potential customers and finding out what their needs are to see if his MossBeGone robot idea has enough value to pursue further. Being an engineer, he often attends customer meetings to explain RoboCo's products and fix problems, but listening for customer needs about a new concept is a skill he hasn't yet developed.

Step 1: Develop a customer persona

To get started, Mark first drafts a customer persona to clarify his target customer as shown in Figure 11.1. He plans to refine this persona after his customer interviews.

MossBeGone Customer Persona

Meet Fred

- Age: 47
- Income: $85,000/yr
- Location: Seattle, WA – Land of trees
- Situation: Owns a single family 3200 sq. ft. home with a shingle roof. He tries to clean his roof every two years. He'll do it himself or hire a roof cleaner if the roof is too bad or he doesn't have time.
- Attitudes: He thinks roof cleaning is a pain and his spouse thinks it's dangerous working on the roof.
- Fred is a *do-it-yourselfer*, but is busy and always looking for easier solutions.

Figure 11.1: Target customer persona

Step 2: Find suitable customers to interview

After Mark creates his persona, he searches for ten customers who might give him accurate insight into their roof cleaning needs. Being in Seattle he knows moss is a big problem because his own mossy roof sparked the idea. However, he also realizes that local customers may not represent the broader market. Mark then finds five interviewees through local friends and family that match his customer profile. And to add more credibility to his data, he finds five early adopters outside of Seattle who might also have a moss problem. He sends off these emails:

> Dear _____,
>
> As our mutual friend, _____, may have told you, I'm working on a project related to helping consumers like yourself keep their roofs clean. I would like to get your thoughts on the topic. Can we find a time to talk in the next few days? Please let me know what works for you.
>
> Mark

Step 3: Develop a needs-first interview guide

To guide his interviews, Mark creates a set of questions following the 4x25 format described above in Figure 11.2. He tests the questions on two of his colleagues to make sure they are clear and will lead to an insightful conversation.

Step 4: Conduct probing interviews

It takes Mark two weeks to complete the interviews. He meets each local potential customer in person, and uses video conferencing for anyone who's not nearby. Ultimately, he's able to talk with six customers of the ten he initially targeted. During those two weeks between scheduled interviews, Mark works on other steps of his 30-Day Action Plan.

Types of Questions

Sample Interview Questions

Explore the Problem (25%)

- Tell me about how you approach cleaning your roof today.
- How big a problem is cleaning your roof? What about moss?
- Walk me through the last time you had your roof cleaned.
- What has been your experience with current products and services?

Explore Customer Needs (25%)

- What frustrates you when you think about cleaning your roof?
- What is most important to you when you look for a solution? Prioritize these.
- What would an ideal solution look like to you?
- What would a device look like that cleaned your roof for you?

Explore Customer Value (25%)

- I'm going to describe a new concept... What is your initial reaction?
- What advantages (disadvantages) do you see over your alternatives today?
- Based on what you said earlier, which aspects meets your needs better?
- On a scale of 1-10, what would make this concept a 10?

Explore Issues and Objections (25%)

- What major concerns would you have about something like this?
- If you just saw an ad for this, what would it need to say to want this product?
- Compared to what you pay today, what is a reasonable price for this?
- So what would you really think about having a robot working on your roof?

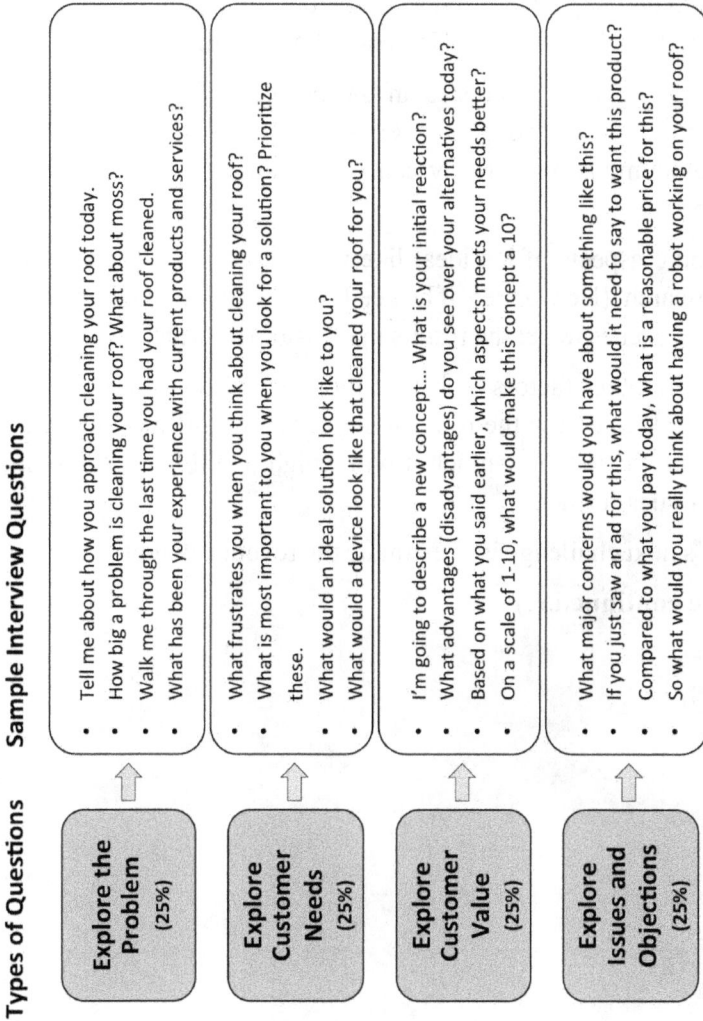

Figure 11.2: Mark's interview guide questions

Step 5: Process results and develop a Customer Needs Pyramid

As shown in Figure 11.3, Mark uses the results of his customer interviews to develop a CNP and a corresponding list of the major objections he heard most often. He concludes the following:

- Customers are interested in the benefits of a roof-cleaning device, but have some serious concerns, such as if an autonomous robot really can clean effectively and if it will be safe.

- Some aspects of his idea, like using chemicals, needs more research. (He wonders if he needs chemicals or could try other approaches to get the results that customers want.)

There are other factors he hadn't thought about that customers find important, like how the robot will get on and off the roof and, if it has to be carried, how much it will weigh. He'll have to investigate these issues further.

Mark's next challenge is to estimate the revenue potential of his idea.

(To be continued...)

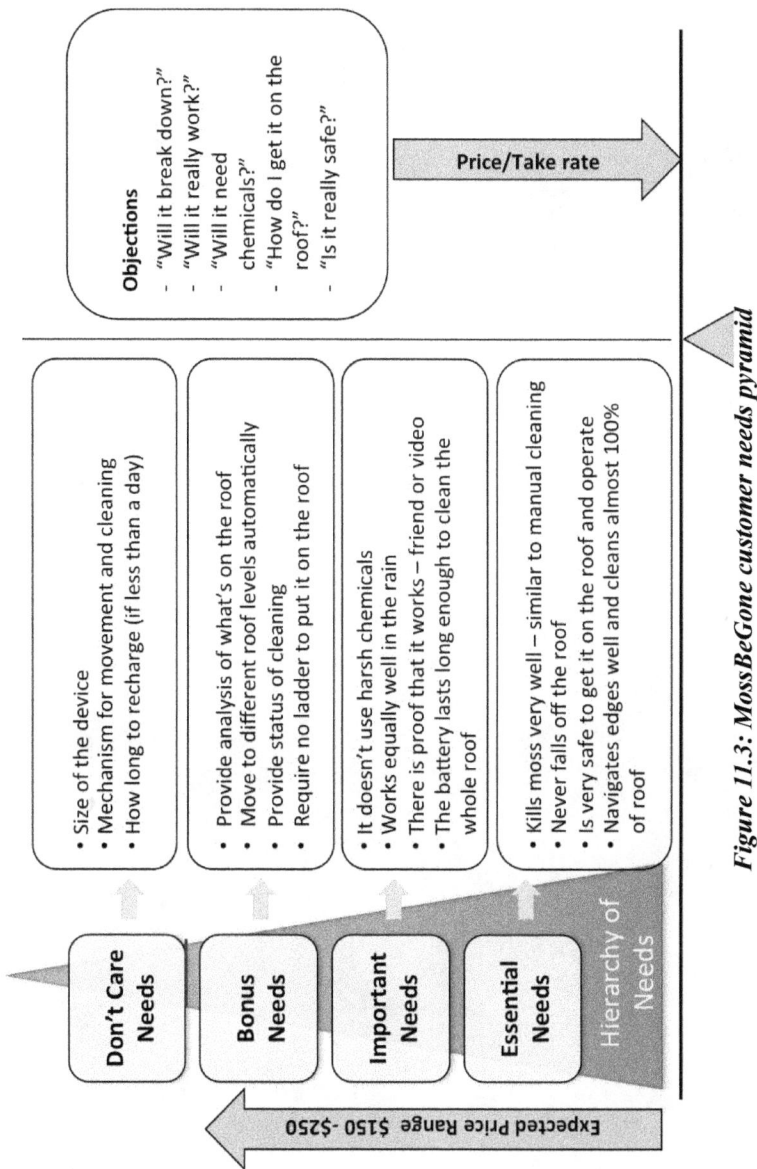

Objections
- "Will it break down?"
- "Will it really work?"
- "Will it need chemicals?"
- "How do I get it on the roof?"
- "Is it really safe?"

Price/Take rate

Don't Care Needs
- Size of the device
- Mechanism for movement and cleaning
- How long to recharge (if less than a day)

Bonus Needs
- Provide analysis of what's on the roof
- Move to different roof levels automatically
- Provide status of cleaning
- Require no ladder to put it on the roof

Important Needs
- It doesn't use harsh chemicals
- Works equally well in the rain
- There is proof that it works – friend or video
- The battery lasts long enough to clean the whole roof

Essential Needs
- Kills moss very well – similar to manual cleaning
- Never falls off the roof
- Is very safe to get it on the roof and operate
- Navigates edges well and cleans almost 100% of roof

Hierarchy of Needs

Expected Price Range $150-$250

Figure 11.3: MossBeGone customer needs pyramid

Chapter 12

ESTIMATE THE REVENUE POTENTIAL

Creating Market And Revenue Forecasts That Persuade Executives

---

Developing a financial forecast for early stage ideas is more art than science.

Savvy Steps To Success

Step 1: Gather Relevant Market Data

Step 2: Quantify Your Target Market

Step 3: Identify Competitors and Determine the Market Situation

Step 4: Estimate a Price and Customer Take Rate

Step 5: Develop a Preliminary Financial Model

Step 6: Obtain Input from Relevant Experts

Step 7: Validate Your Approach with Decision-Makers

Introduction

Now that you've got the wheels in motion understanding the customer value of your idea, it's time to forecast the potential size and revenue of your BIG idea. At this point you don't have enough data yet to develop a full pro forma P&L, but you do have enough information to estimate how big your opportunity could be based on the value of your concept and other market considerations.

In Chapter 8, The Forecasting Challenge, we discussed how an acceptable revenue forecast is really about proving you have a rational approach than any specific numbers. You goal is to convince executives you have a reasonable chance of achieving significant revenue backed up by as much relevant data as you can find in the next 30 days.

This chapter takes you through estimating of size of the potential market and developing a revenue forecast that will be accepted by decision-makers.

Let's get started.

Step 1: Gather Relevant Market Data

If you've followed previous steps, you'll have already gathered a range of data when you identified customers and conducted a rapid market scan. Now you'll need the specific numbers required to develop a revenue forecast. You'll never find perfect numbers in one place nor will you find a perfect piece of information, so you must gather bits of relevant information that, combined, leads to a plausible story.

For now, set a time limit and do the best you can. Focus on using a solid approach to developing a preliminary revenue model with a few key facts. Then get feedback from decision-makers to see if you're on the right track. From their feedback, go back and research the exact data that meets their needs before facing the executive inquisition.

As shown in Figure 12.1, your data needs to explain four levels of information: 1) Total available market, 2) Size of the target market,

3) Size of the accessible market, 4) Number of customers you believe will buy your concept (the customer take rate).

To develop these estimates, look for these types of data:

- Information on industry market sizes, segments, and trends
- The number of customers who have purchased similar products or services
- Trends in customer needs, behavior, spending, and market dynamics
- Companies and products that currently or plan to offer similar and related products
- Prices of similar and related products and services
- The state of technology, costs, and supporting infrastructure available to develop or deliver your product

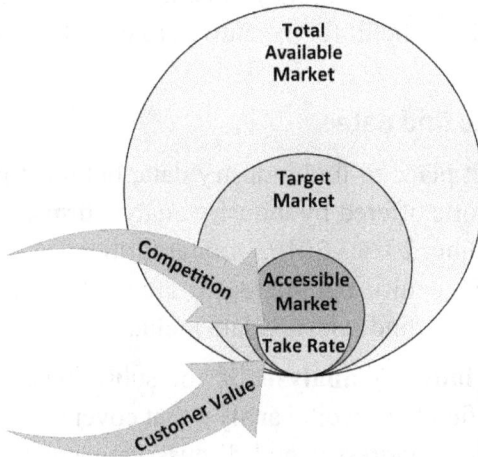

Figure 12.1: A model for estimating revenue

For brand new products or services where directly-related information is not available, try to find data that shows how customers currently solve the problem you've identified. For example, if you were researching the first mechanized elevator, you wouldn't say, "There is currently no data available on this exciting new market!" Instead, you'd look at what other companies are doing to solve the problem (of going up and down) with stairs, ramps, and manual pulleys, and you'd present that data related to those solutions.

Trends provide the wind behind your idea

Don't forget to look at information on the various market trends related to your idea. Data on trends are critical bits of information, and most executives are studying them too as the basis for their investment decisions. They know it's easier to make profit in categories that are growing and where customers are more likely to spend money in the future. This lowers the risk of investing in new opportunities because your company will have additional options to change directions and follow a trend when things don't go right. The popular term *pivot* describes the process of changing plans as new information becomes available over time. If you can't show you'll be entering a market where money is being spent on similar products, then it will be difficult for executives to take the risk.

Typical places to find data

There is no secret place to find industry data, but most people start and end with reports offered by industry analyst firms, such as "US Healthcare Spending: 2014 –2019." Going further than this can take time and a little ingenuity, but the results are well worth the investment. Other places to find relevant data include:

Direct from industry analysts. If you subscribe to an industry data service, find the specific analyst that covers the market segments your idea addresses and discuss your needs. They may have data you can't find in a report, be willing to share their approach, or point you in the right direction.

Venture capital investment. One way to provide evidence of an emerging market is to identify the trends in early stage investment of emerging technologies, including the companies and types of products and services being funded by venture capitalists.

Industry associations. Most industries have an association that collects data as a service to their industry members. If you're researching a product that is outside the industry you're currently in, attending a tradeshow is another way to get familiar with the new industry and their latest products, technologies, competitors and key players.

Industry consultants. Find a consultant who works in the area you're researching. It may cost a few hundred dollars for an hour of their time, but if they point you to the right reports and share their expert opinion, you'll save days of hunting. Many consultants also provide free advice with the hope of getting future business with your company.

Once you've obtained a range of data related to your potential market, you can proceed to develop a preliminary revenue model.

Step 2: Quantify Your Target Market

Forecasting revenue typically starts by estimating all of the potential customers for a product or service as well as the possible revenue if every potential customer purchased the product. This is often referred to as the Total Available Market (TAM). Getting an industry estimate for a broadly-defined TAM is pretty easy. If you're investigating a new idea for a 3D printer component, you could read an industry report and conclude, "The overall market for 3D printers will be $2.4 billion in 2015." However, this is where a lot of the industry data ends, so many innovators stop here then jump directly to a percentage of this market to figure a revenue estimate. In addition, a TAM estimate provides little evidence to explain exactly who and how many people may buy your new type of 3D printer component. So remember, it's ok to start with this broad number, but you must also explain the subsets such as 3D printers for specific applications, different price points, different customer needs, etc.

A savvy innovator knows how critical it is to develop a solid argument on the specific type of customer—the target market, and why elements like the customer persona and customer interviews are critical as well. Without these insights, you're just guessing which segments of a broad market are likely to be customers. Guessing without solid backup data simply exposes your opinion to someone else's.

Step 3: Identify Competitors And Determine The Market Situation

Your potential revenue will be largely based on the strength of current and potential competitors. It's always difficult to quantify the impact of competition in a revenue model. Statements like, "our price will be lower," "our brand will drive sales," and "we'll offer more features," are all weak arguments that don't provide the data necessary to support a funding decision. Many innovators say, "We don't have any competition!" Every decision-maker knows this isn't true. Even if no other company offers a similar product or service, there is always competition. It just depends on how you define competition. Sometimes the most powerful competitor may be the inertia of a fixed pattern in your potential customers' brain that says they don't need your new product.

To make a credible argument about the unique value your idea brings to your target market, consider three types of competitors:

Direct Competitors. The most obvious competitors are those with products or services that solve the same problem you are solving using a similar approach. In established markets these are easy to find and explain.

Indirect Competitors. These are less easy to spot, but will still have a significant impact on your market potential. Companies and products may be solving a similar problem to what your idea solves, but with a different type of product or service. Indirect competitors may also be solving only part of the problem you're solving.

Non-Competitors. Somehow customers have found ways to solve their problem by not purchasing anything, so they must be convinced of buying something to solve their problem (assuming they even recognize the problem). For cutting-edge products and services, non-competition is often your biggest competitor.

For a new concept, you may believe that no direct competition is a good thing. However, executives often believe there is more risk in having no direct competition than in having many direct competitors. They are right to be skeptical because often no competition means there are few potential customers.

If you believe there is no competition to your new concept, do this simple test...ask yourself, "Do customers spend money on solving this problem today?" If you can answer 'yes' and support your answer with data, then you should be able to explain your revenue model and show your market potential. The inventors of microwave ovens, mainframe computers, and even the telephone all struggled to answer this simple question. Certainly people spent money on heating food, processing information, and communicating with others, but translating this to the market potential for each of these radical ideas would have been difficult. Some inventors were able to make this translation and became famously successful. Others failed to be recognized for their brilliance...which one do you want to be?

Keep competitor analysis simple and relevant

If you have many competitors, you can spend a massive amount of time surfing the web and finding loads of competing companies and products. Be careful! It's not your goal right now to list all the details on every competitor. To be efficient while researching, you need a focus, such as how well the competition is meeting specific customer needs based on the customer insight you learned earlier.

Identify how other companies are solving the same problem you are, how they're positioned in the market, and what their strengths and marketing efforts are. If a competitor has a history of high market share or launching innovative products, treat them as a serious threat and downgrade your revenue estimates. Even smaller competitors can't be ignored; you must consider their momentum and possible strength in the future. By all means, don't say things like, "The competition will be too slow to respond" or "Our product will always be better than the competition." A savvy innovator knows ALL competition must be taken seriously.

Determine the market situation your concept is entering

To develop a believable forecast, you must be able to explain the market situation as discussed in Chapter 8. If you're entering a mature market and need to "steal the market" from strong competition, your forecasts and data must reflect this situation. The same is true if you need to find, build, or energize the market. Part of your data gathering must be to identify the trends, strength of competition, and the newness of your concept to determine the market situation you'll be entering.

Step 4: Estimate A Price And Customer Take Rate

There is no variable that has the largest impact on a revenue forecast or generates the most disbelief as your estimate of the customer take rate. The challenge with estimating a take rate is that it is based on everything: from the value of your product, to the strength of the competition, to the price and the quality of sales and marketing tactics. A take rate estimate can never be proven without doing extensive market trials or other advanced market research. If you've done your customer work well, you should be able to convince executives that a sizable number of potential customers will actually purchase your product. To improve the credibility of your take rate estimate, carefully research and answer these three questions:

> **How many customers can access the product?** With physical products, you usually access customers by building a distribution channel such as partnering with distributors, value-added reseller, or retailers. However, ecommerce makes it easy to assume that everyone in a target market will have immediate access to a new product or service. However, just because a product is available to the world, doesn't mean customers will

find it. Too many innovators assume that customers will just mysteriously show up to a website. In your proposal, you'll need to explain how exactly you will find customers and how they will find you.

Is it easy for customers to see the benefits? The take rate of a new product is highly based on how quickly customers see and appreciate the benefits. Complex products that seem difficult to use or have obscure benefits will slow down customer adoption. If people can see other people enjoying the benefits or they talk with others about it, then adoption will accelerate. This is why apps like Angry Birds and Instagram had dramatically fast adoption rates. Both were extremely easy to understand and share. Your take rate must first prove that potential customers can quickly see the value in your concept. Some may see the value right away. Others may need to see proof that it works. And others may need to try it for themselves before they believe anything.

How difficult is it to create demand at the expected price? This question builds on the last two. It's not enough to get in front of customers and have them appreciate the value of your concept. You must go further to show that potential customers are interested enough to actually purchase a product at a specific price. While it may seem early to estimate a price point, it's impossible to estimate a take rate without considering the expected price of your concept. Since price is closely related to perceived customer value, the higher you set your price relative to that value, the lower the demand you can expect. A good example was the launch of the Segway personal transportation vehicle. The company estimated large customer demand, but its price point (nearly $5,000) became a major obstacle in customer adoption for a product that did not meet essential needs.

FORECASTING SCRIPT

Once you've got actual revenue estimates, you can use this script when presenting a forecast:

"Based on my current data, I believe the target market for (your concept) is the set of customers who (specific characteristic or need) or about X% of the (broader market). Based on five in-depth customer discussions, (key benefit) is highly valued by the majority of these customers. These customers also understood the benefits easily. My forecast is based on accessing customers through (means of accessing customers) and being able to demonstrate the benefits using (types of marketing activities). We will continue to learn more with (this plan)."

Step 5: Develop A Preliminary Financial Model

To summarize, for each period of time you are forecasting, revenue is a result of this formula:

Revenue = Size of Accessible Target Market x Take Rate x Price

There is no one correct format, but most revenue models start with the equation above and includes:

- An estimate of the target market size
- An estimate of how many customers can access the product
- The rate that customers will purchase the concept (customer take rate)
- An estimated price

You may need to modify the basic equation to reflect your opportunity's unique situation and business model. Include, for example, distribution partners like distributors or retailers. You might also base your revenue model on repeat purchases, giveaways to sell other services, etc. Whatever model you create, it must be simple, accurate and believable. Remember to avoid the garbage-in/garbage-out forecast. A complex model with lots of variables and faulty data has little chance of earning support.

Step 6: Obtain Input From Relevant Experts

You may have convinced yourself that your revenue model is good enough to gain executive approval. But will it pass the inquisition? In most opportunity environments, a believable revenue estimate has been approved by knowledgeable experts who will validate your data and approach. Experts might be internal analysts who are respected for developing forecasts or possibly outside consultants from the industry you're investigating.

Validating with the sales team and channel partners

Depending on your idea and company environment, there is one source of input you shouldn't ignore: input from your sales team and/or potential channel partners. If your idea is for next-generation products, the sales team should be able to help since they know current customers and products. But regardless of the product and despite what the sales team may or may not know about the market, get input from them anyway because they have a lot of power in the company. They're viewed as experts at acquiring customers and they may be on the hook for achieving sales forecasts (even for completely new types of products). Often a decision-maker asks someone in sales, "What do you think? Could you sell this? Are these volumes reasonable?" If the sales team wasn't involved with the forecasting process, then the response will not be favorable to you.

Sales forecasts are not gospel

Involving the sales team in forecasting doesn't mean you accept their forecast. The one you developed using the market-based revenue model in the previous step will likely be dramatically different than what the sales team gives you. They often forecast low since they generally have a limited knowledge of the market or they believe they are committing themselves to a high sales target once the product is out. On the other hand, they could give you a wildly optimistic forecast because they want to see the product built for a key customer.

Whatever input you get from the sales team, make them part of the process and respect their input. And if necessary, figure out why their forecast is different from yours and how to use this information.

Step 7: Validate Your Approach With Decision-Makers

The last and most important step when developing revenue forecasts is to gain support from decision-makers and other influential executives. They may not be any more knowledgeable about the market or be able to develop a more accurate forecast than what you've created, but they are the customers for your opportunity proposal, and your success is based on their acceptance. Because of the inherent controversy revenue forecasts create, seek out the most influential executive you can find to validate your data, approach, and results. This may be a senior marketing executive, CFO, or even the CEO in some cases.

When you do meet with them start by saying, "Here is what I'm trying to forecast. This is the data I found and here is my approach. What do you think?" You could also ask:

- What major objections do you see with this approach?
- What is the one factor you need me to address if you had to accept this forecast?
- Do you have any examples that have worked in the past?
- Is there someone I should work with to get this right?
- Can I review the results with you when I am close to being done?

Afterwards, keep working on it until an executive will stand up and proclaim, "I reviewed this forecast. It's the right approach and has acceptable numbers for this stage of analysis."

Sometimes they'll acknowledge that they don't know enough about the market to even tell if you're on the right track. They might want an industry expert to review (or even develop) your revenue model. This is actually a good thing. It shows they care enough about your idea to invest in outside expertise.

CHAPTER 12 REVIEW—ESTIMATE THE REVENUE POTENTIAL

You'll find that developing market and revenue forecasts will be the most challenging activities as an innovator. But keep at it and work the steps in this chapter. Focus your efforts on developing a rational approach with enough data to convince executives that your idea has the potential of meeting your company's revenue and other strategic goals. And wherever you need the guidance or sources of data, seek them out to add further credibility to your forecasts.

Savvy Success Timetable

Times will vary by project and executives' needs, but a savvy innovator should use a balanced approach to address every aspect of an opportunity. As a guideline, here are the estimated hours you should spend following these steps of your 30-Day Action Plan to determine revenue potential:

Step 1: Gather relevant market data	8 hours
Step 2: Quantify your target market	2 hours
Step 3: Identify competitors and determine the market situation	4 hours
Step 4: Estimate a price and customer take rate	4 hours
Step 5: Develop a preliminary financial model	2 hours
Step 6: Get input from relevant experts	2 hours
Step 7: Validate your approach with decision-makers	2 hours
Total:	**24 hours**

MossBeGone
Mark's 30-Day Action Plan (continued)

Objective 3: Estimate The Revenue Potential

Now that Mark has a solid understanding of his MossBeGone autonomous roof-cleaning robot concept, he's ready to estimate its revenue potential. He knows that no matter what, his numbers will create controversy, so he focuses on getting early buy-in from key executive influencers.

Step 1: Gather relevant market data

Mark identifies a range of relevant (and not so relevant) data from a variety of sources. This includes the two reports he purchased online earlier. From these sources, Mark gathers the following:

- Industry sales for consumer robots are currently about $500 million/year, growing at 14% annually and expected hit $1.5 billion by 2019. By comparison, industrial robots are currently an $11 billion market but are declining. Robotic vacuum cleaners are the major portion of consumer sales today with some sources estimating this category makes up two-thirds of the consumer robotic market.

- There's a wide range of robots gaining traction with consumers such as pool cleaners and lawn mowers, but sales estimates are hard to find.

- Most roofs in America have a pitch (steepness) of between 15 degrees and 30 degrees, and about 60% of roofs are made of shingles.

- Based on data for consumer vacuum cleaners, about 1% of homes purchase robotic vacuum cleaners.

- There are about 76 million single family homes in the United States that aren't condominiums or apartments.

Mark can't find any current industry information on how much money homeowners spend cleaning their roofs. However, he does call several roof-cleaning services and uses the results of his customer interviews to develop an estimate. The quotes vary widely from $500 to $2,500 for a 2,500 sq. ft home.

Mark finds no exact products like his, but does find a mechanical device that cleans roofs. It sells for $400. But an operator has to stand at the peak of the roof and use a rope to control the device.

As Mark prepares and documents his data, he carefully notes the sources where he gets the information.

He also notes that while this information begins to indicate that a robotic roof cleaner might be a sizable market that could grow, none of it directly indicates the number of people that might purchase a MossBeGone robot. Therefore, he'll need to weave this tangential market information into a preliminary revenue forecast.

Step 2: Quantify your target market

Obviously Mark can't just target people who own homes, so he has to narrow the market to a specific set of customers who would most likely need his robot and then quantify this smaller target market. Mark can use several of the following customer characteristics to break the broad market into smaller segments and define a more rational target market:

- Homeowners in geographic regions that have significant moss problems or other roof cleaning needs.
- Homeowners with roofs that can be easily navigated by a robot; because of the pitch, complexity of the roof structure, or type of material.
- Homeowners with larger roofs that would take longer to clean.
- Homeowners with more expensive homes where the owner will be willing to spend money on a cleaning device.

After looking at his options, Mark decides to develop a TAM by estimating the number of homes in regions that likely have moss problems and of those, the subset of homes with the type of roof a robot could navigate. Mark theorizes his first target market is the 70% of homeowners who have shingle roofs and live in 35% of the United States where moisture and trees create moss problems (mostly northern states). This is approximately 18 million homeowners (as shown in Figure 12.3).

Mark realizes this target market is just a hypothesis right now. He could slice the market many ways, but with his current data, any estimate will be met with skepticism. And that's OK. He primarily needs to answer, "Is there a big enough market for a moss robot to be highly profitable?" Mark knows he can't just point to a large number of homeowners and say, "I think that 5% of these potential customers are good candidates for the MossBeGone robot." It's more believable to point to a specific segment of the market and say, "This is the overall market and some possible segments. Based on the benefits a moss-cleaning robot will provide, the first target market I'm researching is homeowners within these regions with this type of roof. Here is why…." Later on Mark knows he can change this initial hypothesis as he learns more.

Another challenge was trying to estimate how many customers could access the product. He assumes MossBeGone will be available to anyone online, but anticipates most sales will come from the retail channel. He broadly estimates the product will be available in a small number of stores initially (providing access to 5% of the target market), then grow to major retailers over several years before finally being available to 80% of the target market.

Step 3: Identify competitors and determine the market situation

Mark summarizes the competition as follows:

- Direct competitors—There are no direct competitors offering a similar device.

- Indirect competitors—There are a wide range of products and services from the roof-cleaning device that requires a manual operator to many roof-cleaning services. He concludes the roof-cleaning services are the biggest indirect competitors.

- Non-competition—Mark assumes that a large number of homeowners do very little to clean their roofs. Many regions don't have moss problems, so he'd be unlikely to convince people of buying his robot if they don't already use a product or service in the indirect competition category.

To help his audience visualize this data, Mark makes a two-dimensional matrix (shown in Figure 12.2) using two of the primary customer needs: affordability and minimal personal labor. He puts the different types of competition in each quadrant to summarize the desired options and to indicate where MossBeGone would be positioned in the market.

Affordable

Do-it-yourself
($50 in supplies and
valuable time)

MossBeGone
($300-$500)

High Personal Labor

Low Personal Labor

Do-it-yourself
(Cost of injury!)

**Roof
Cleaning Services**
($500-$2500)

Expensive

Figure 12.2: MossBeGone competitive landscape

Step 4: Estimate a price and customer take rate

When Mark tries to estimate the potential customer take rate, he really hits a wall. He knows that if he estimates too high, it won't be believable. If he estimates low, then the numbers look too small. Based on his customer interviews and discussions with retailers, Mark thinks the prospective benefits of MossBeGone are high, but customers will have to be convinced it works. Of the six customers he interviewed, two of them said they'd buy right away. But 33% is too high a number to justify on such as small sample. He decides to show a progression to the executives to get their reaction based on the consumer robotic vacuum cleaner purchase rates. As a starting point, he estimates an initial take rate at 1% and rising to 2% over the seven-year forecast period.

For the price estimate, Mark doesn't have a lot of information yet, but does have enough customer insight for the purpose of estimating revenue. He believes the retail price can start high at $500 for a short period at launch and then be reduced to a $300 price point over two to three years.

Mark learns from his sales team to assume that retail distribution partners will want a 35% margin. This means MossBeGone would need an initial wholesale price starting at $325 per unit and come down to a wholesale price of $195 to hit the $300 retail price. Mark immediately realizes these price points would create a margin challenge for his executives. If he shows a gross margin of 50%, which is what his executives are accustomed to achieving with industrial robots, then his total production cost would need to be less than $100/unit within 2-3 years. Mark wonders if this is even possible? Clearly both the price and costs will have to be validated as part of his MRA.

Step 5: Develop a preliminary financial model

Using his new estimates, Mark creates a preliminary revenue model as shown in Figure 12.3. In addition to retail sales through distribution, he also assumes there will be enough direct sales through RoboCo's website to count as a significant revenue component in the model. Estimating worldwide sales is another challenge. The United States constitutes about one-third of the global GDP, but he has no idea if they can find international partners or if a roof-cleaning robot will even be valuable internationally. To facilitate discussions, Mark adds a line-item for international sales at a level of 25% of US sales.

Because Mark can make these numbers say anything he wants them to (and his executives won't believe them), he realizes he needs to get them to believe in his story and see the same potential in this opportunity that he does.

Step 6: Obtain input from relevant experts

Now Mark decides to leverage the expertise and power of the sales team at RoboCo. Although they had little understanding of the consumer robot markets, one sales team member did have some experience selling to hardware stores. She helps Mark understand how retail buyers make decisions on carrying new products and how they manage inventory and customer returns of products similar to the MossBeGone. This input will at least give Mark some credibility until he can find someone with direct expertise. He'll also have to address this in his MRA and subsequent phases.

Step 7: Validate your approach with decision-makers

With his revenue model in hand, Mark sets up another meeting with Tony, the CFO. Tony generally agrees on his forecasting approach, but as expected, is skeptical of the numbers, primarily concerning the market size. Tony asks, "If you can prove there are enough roofs out there that have this problem and we can actually solve it, I'll be willing to support developing a prototype." This is all Mark needed to hear. He sets about talking with people in the home construction industry to get his numbers accurate and big enough to be interesting to Tony.

Mark's next challenge is to develop a preliminary go-to-market plan that can convince executives that his concept can become reality both as a product and as a market success.

(To be continued...)

MossBeGone 7-Year Revenue Forecasts

	Investment Period		Profit Period				
	Year 1	Year 2	Year 3	Year 4	Year 5	Year 6	Year 7
US Single Family Homes	76,000,000	76,000,000	76,000,000	76,000,000	76,000,000	76,000,000	76,000,000
Target Regions (35%)	26,600,000	26,600,000	26,600,000	26,600,000	26,600,000	26,600,000	26,600,000
Target Roof Types (70%) (TAM)	18,620,000	18,620,000	18,620,000	18,620,000	18,620,000	18,620,000	18,620,000
% of retail store coverage	n/a	5%	25%	50%	65%	80%	80%
Total Accessible Market	n/a	931,000	4,655,000	9,310,000	12,103,000	14,896,000	14,896,000
Take Rate Estimate	n/a	1.0%	1.5%	2.0%	2.0%	2.0%	2.0%
US Retail Sales (units)	n/a	9,310	69,825	186,200	242,060	297,920	297,920
Internet Sales (+20% of US retail)	n/a	1,862	13,965	37,240	48,412	59,584	59,584
Global (25% of US retail)	n/a	2,328	34,913	93,100	121,030	148,960	148,960
Total Unit Sales	n/a	13,500	118,703	316,540	411,502	506,464	506,464
Revenue							
Est. Retail Price	n/a	$500	$400	$350	$300	$300	$300
Wholesale Price (Retail Margin ~35%)	n/a	$325	$260	$228	$195	$195	$195
Units	n/a	13,500	118,703	316,540	411,502	506,464	506,464
Total Revenue	$0	$4,387,338	$30,862,650	$72,012,850	$80,242,890	$98,760,480	$98,760,480

Figure 12.3: MossBeGone revenue estimates

Chapter 13

DEVELOP A PRELIMINARY GO-TO-MARKET PLAN

Address Product Development, Marketing, Sales, And Other Operations Tactics

---

Ideas are a powerful start to innovation, but tactical plans lead to the results.

Savvy Steps To Success

Step 1: Create a Unique Selling Proposition

Step 2: Develop a Preliminary Product Roadmap

Step 3: Clarify Your Enabling Capabilities

Step 4: Develop a Preliminary R&D Plan

Step 5: Describe Key Sales and Channel Tactics

Step 6: Describe Key Marketing Tactics

Step 7: Describe Your Risk Mitigation Plan

Introduction

Up to this point, you've largely been in the imaginary world of hypothetical concepts, customers, and revenue. But it's time to transform your idea into a fundable opportunity and think about the specific tactics necessary to realize real-world results. While it may seem early yet to plan specific tactics for developing, selling, marketing, and producing your idea, few executives will accept the risk associated with a new opportunity without some idea of how to execute it.

In this chapter, we'll explore these activities from specific research to targeted marketing promotions that make your plan real. You'll be helping decision-makers visualize real people, doing real activities, and spending real money on your idea.

Let's get started.

Step 1: Create A Unique Selling Proposition

In marketing there's a concept called a Unique Selling Proposition (USP). A USP defines the primary basis used to market and sell a new concept. This is important because every aspect of your tactical plan will support whatever you select as your USP. Remember the 20-second opportunity pitch discussed earlier? It's like a USP to get executives interested in your idea. The concepts are similar, but now we think from the buying customer's perspective to convince executives you know precisely what is most important to your target market.

As an example, say you have a new type of teeth whitening product. Your Customer Needs Pyramid reveals that potential customers have two essential needs: 1) a consistent white they want to achieve and 2) a rapid whitening. From your competitive homework, you've learned that your new idea performs better in the consistency of whiteness because it works on the full set of teeth with no splotchy results compared with whitening strips available today. This advantage will be your USP.

A general USP formula looks like this:

"For customers who (customer profile and problem), (concept) provides (attribute that solves the problem or fills the primary need). Compared to alternatives our offering provides (key advantage)."

Now, fill in the blanks using the teeth whitening product:

"For customers who are looking to obtain a bright, healthy-looking smile from a home solution, the teeth whitening product I'm proposing provides a better teeth whitening experience. Compared with current teeth whitening solutions, this new product whitens a person's complete set of teeth and guarantees the most consistent level of whitening available."

At this point your USP might be a hypothesis that you'll test in later phases of development. But it's good to have one now because it shows you understand the primary customer and your main advantage. Once you've got a USP, your product, marketing, price, and distribution channels can be developed around this proposition. This creates a clear, cohesive beginning to a story that executives can easily follow.

Step 2: Develop A Preliminary Product Roadmap

You probably have an extremely clear vision of your concept that includes a wide range of exciting features and functions, especially if you listened well to your customers. However, it's likely your vision includes far more features and attributes necessary to enter the market successfully. If so, the product will be expensive to produce and it may take too long to develop, which will make achieving profit difficult. Launching a complex product creates a high degree of both development risk and market risk. A decision to enter the market with an initial product that costs $3 million to develop is very different than a spending $40 million to enter a mature market.

Consequently, you don't need, nor should you spend time developing, a fully defined product for your initial opportunity proposal. But you do need to clearly define the product attributes that will create the most value for customers to be successful in the market. And this includes anticipating market conditions when your solution is likely to hit the market.

Ensure that your concept is viable

Eric Ries, author of *The Lean Startup*, has built a popular concept in the startup and Agile development communities known as the Minimum Viable Product (MVP). An MVP is a set of product attributes that are just good enough to be "viable" in the market. This allows a company to enter the market quickly with a valuable, but not over-featured, product or service. Once a product is in the market, real-world data can be used to learn how to make it more successful.

The emphasis on an MVP should be the word *viable*. Viable is relative to the market situation. If your concept is early in the market with few, if any, direct competitors, your product can be successful with a minimal functionality that creates a unique offering. However, if your concept is entering a more mature market with serious competitors, then your MVP must be more detailed and more differentiated. It must provide additional value over the competition. As you think about your MVP when presenting your product roadmap, you must be able to answer the question, "What proves this set of attributes will actually be viable in the market?"

At this writing, Blackberry and Microsoft/Nokia are struggling to gain market share with their recent smartphones. If they had launched early in the smartphone maturity curve, their products would have looked outstanding compared to current products. But, the market was already mature with several generations of iPhones and Android-based competitors that provided great value and full application ecosystems. As mobile phone companies know, when entering a mature market, an MVP must be defined with more innovative and valuable differentiators.

A clear roadmap starts with a clear vision

Figure 13.1 shows the general approach you should take to developing a product roadmap. A full vision usually takes more than one product release. This may not be the case for a simple consumer product like a board game or kitchen utensil, but even with these, the same approach should be considered. For example, the popular board game, *The Settlers of Catan*, started as a basic game of gathering resources and creating settlements. As it became successful, the company added extensions for new worlds and electronic versions.

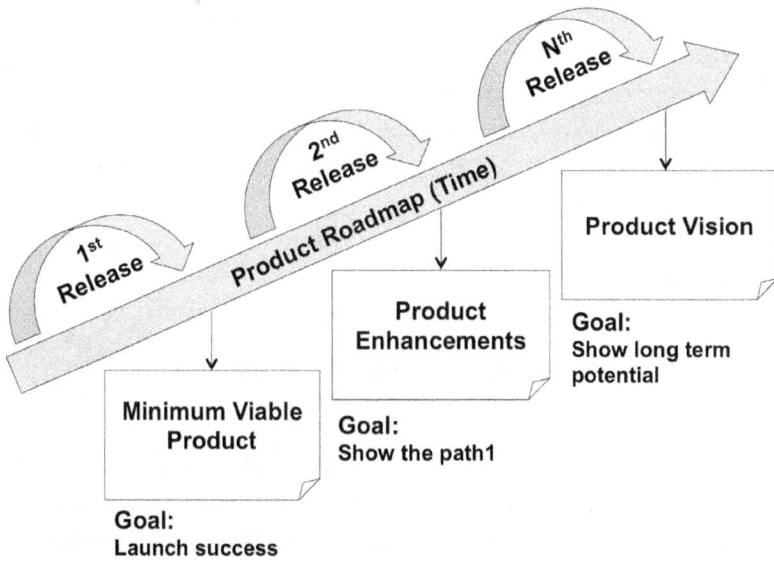

Figure 13.1: Roadmaps from MVP to vision

If you have initial market success, there will always be more resources to make your product better and realize your vision. Success starts by clearly articulating your initial product and then drawing a logical path to your vision. The vision should be exciting—the initial product must be viable.

Step 3: Clarify Your Enabling Capabilities

Any successful innovation is often the culmination of key trends in technology or other enabling capabilities that were combined to address a pressing customer need. Look at our history: the enabling capabilities of increased computing power, lower air transportation costs, and logistics software made FedEx possible; new capabilities in hard drive memory density and MPEG2 encoding made the digital video recorder Tivo possible; and web browsers and cloud computing has made services like Salesforce.com possible.

If your opportunity is based on one or more key technology trends (probably even sparking your idea in the first place), everyone will have an opinion about the capabilities of those trends, what they mean for the future, and how to apply them to your concept. A savvy innovator must identify these emerging capabilities and understand how they should (or shouldn't) be applied to a new opportunity. When you face the executive inquisition, you'll need to debate your approach and its relative merits against the collective experience and opinions that are in the room with you. So think through these enabling capabilities before you make the big pitch for funding.

Should Your Mousetrap Be More Radical?

You'll need to answer the question, "What new technology or enabling capability will give us a clear market advantage?" The answer is relative to the market situation you are entering. We've all heard the saying by Ralph Waldo Emerson that you can, "Build a better mousetrap, and the world will beat a path to your door." Let's look at how you might innovate on the lowly mousetrap based on two different market situations:

You're already in the mouse trap business. You could build a superior mouse trap that is 10% more effective than existing designs and would probably be very successful. Simply replace your existing product in the market or offer another premium product at a higher price. Leveraging your current brand and sales channels would make your development investment very low risk and easy for management to support. This is a classic example of 'incremental innovation' that's relatively easy to justify and can be highly profitable.

You don't currently sell mousetraps and you want to enter an established market that already has serious competition. You won't likely succeed by offering a solution that is 10% or even 20% better than existing solutions. Mousetrap customers live in a dense and noisy market filled with competing products, brands, and pre-conceived notions about mousetraps. A slightly better product will only receive a yawn from the marketplace as well as executives. Therefore, you need a more radical innovation—with a serious wow factor—that can demonstrate a 30% or even 100% better option for gaining rapid attention in the crowded world of mousetraps. To create this significant advantage, you'll need to think about enabling capabilities, such as mouse detection sensors, or a mobile application that notifies the customer that the trap is full, or perhaps a technology that makes a more mouse-friendly trap for customers who worry about hurting the poor mouse. However you approach your new mousetrap, you'll need to make a leap in mouse eradication that only one or more new enabling capabilities will provide.

It's tempting to add more and more capabilities to your concept. But be cautious because this can create an expensive, complicated or delayed product with a lot of whiz-bang attributes that customers just don't care about. On the other hand, if you haven't identified at least one enabling trend that gives you a market advantage, you run the risk of creating a 'me too' product won't get the attention of customers or executives.

Step 4: Develop A Preliminary R&D Plan

With your product roadmap in hand, it's time to create a product development plan that matches the roadmap. You don't necessarily need a detailed development schedule (unless executives ask for one), but you should list key steps along with the quantity and types of resources needed to execute the development effort—at least for getting to the initial MVP. Whenever you estimate development efforts, there'll always be tradeoffs between time, resources, and product attributes. A savvy innovator focuses on creating three things:

- A magnitude of time and resources based on your initial product release as well as clear milestones.

- An estimate of R&D expenses.

- A list of any risks associated with the skills required.

Most companies know that a development plan at this stage will be very preliminary. But executives still need an estimate of the time, resources, and risk involved. If you're a technologist, creating your R&D plan may be easy, but you may also be tempted to get too detailed. If you're a marketer or in sales, obtaining rational estimates may be difficult since you may not have a clear understanding of the technology, how long development should take, or the skills and expenses needed for development.

OVERESTIMATING ESTIMATES

Paul, a skilled marketer and innovator for company that focused on social networking applications, had a vision for a more visual, real-time experience that allowed audio, text and images to zoom in and out based on a consumer's level of engagement. The interaction model was similar to the virtual wall experience Tom Cruise used in the movie "Minority Report." Paul had developed a set of compelling drawings and wireframes to prototype how it might work. He took these to customer interviews and discovered that customers loved the new experience. He then asked his internal development team, "Can you guys give me an estimate on what it might take to build this?"

Paul tried to explain that he only needed a ballpark estimate to build a business case and help him determine how difficult it might be to develop and whether the team could actually develop it. However, what his team heard was, "This is a real product and we'll be responsible for any cost and schedule estimates we provide."

It took some coaxing but eventually the development team

provided an estimate. Assuming the prototypes included required features and functionality, the team presented a spreadsheet that included every feature and a "worse-case" development estimate. The result was a very high development cost of $3.7 million that would take 15 months to develop. While Paul was not a technologist, he knew enough to know the team was sandbagging.

In most situations, people don't want to underestimate a development effort for fear that once the estimate is on paper it gets built into their future objectives and subsequently their success (or failure). So they often "sandbag" by showing numbers that lower their risk. Savvy innovators understand this phenomenon and work hard to assure people that it's OK to provide ballpark estimates that will be refined later and that they won't be held accountable for preliminary estimates.

R&D forecasts also need executive selling efforts

Just as your revenue forecast will never be totally accurate, neither will your R&D estimates. You know it. I know it. Executives know it. You must use the same path in gaining support for your development estimates as you did for your revenue estimates—decide an approach and then work with respected executives to understand how to make your estimates acceptable. Some concepts will apply existing technology in new ways. Others will require completely new technology. For those ideas, part of your proposal might show how you intend to identify technology partners, use vendors to outsource R&D, or hire technical leaders with new skills. The specific tactics of each strategy may not be necessary, but explaining your overall approach to success is.

Your R&D estimate is not what you are asking for

The resources you show in your development plan are not the resources you'll be requesting when you present your opportunity proposal.

It's necessary to show the magnitude of resources needed to fully develop your opportunity so you can discuss profit and ROI with executives, but your real funding request is your MRA. Your MRA should only include first-stage resources—what you need to reach near-term milestones. These may include simple prototyping, technical investigation, or no R&D milestone at all.

Never negotiate attributes needed for a viable product

If you've ever developed a new product, you've probably been forced to make necessary trade-offs between time, resources, and features during the product development process. This is normal and expected. Even companies with massive R&D budgets like Apple, Google, Microsoft, and 3M, all have to make day-to-day tradeoffs when deciding whether they have the resources to develop a specific technology or feature (like if a new phone needs a carbon-fiber case or if hardened plastic is good enough). I personally would love a cutting edge website with exciting features and interactive video, but I know I have to make resource tradeoffs and live with what I can afford.

The question you must always ask yourself is, "Who will be the one that determines if this concept is good enough and how will these decisions be made?" Maybe you don't need to decide now during this 30-day period, but I promise it will come up soon. The question usually comes up when you ask a development team to make estimates on developing your concept, so be prepared. You may not have to worry about this problem for months yet, but you are laying the foundation for the discussion now, in the earliest stages of your thinking.

Having clarity on the real priorities of your concept and where to push for real innovation will be critical to your success. To prepare for tough tradeoff decisions, be crystal clear on your customers' needs and exactly what will make your concept viable in the market—and don't negotiate on these critical attributes! If you're not clear and don't have data to support your clarity, others will gladly make a decision based on what they think is important to customers, which will ultimately determine your own success.

Step 5: Describe Key Sales And Channel Tactics

You've had to think about sales tactics earlier when you developed a revenue forecast. Now think more about the specifics. Executives will always question your sales and distribution plan with, "Are you planning to use the existing sales team, ship products directly to consumers or use a distribution network such as a retailer?" Bottom line —executives want to visualize your concept being sold. If you need a new sales team, you'll have to describe the type of sales person you need, how he or she will fit into your organization, and how you will recruit. If you're planning to sell directly to customers via a website, you'll have to explain how you'll take orders, handle customer service issues and manage product returns.

To create a cohesive story, your sales and channel tactics must match your overall strategy, with no discrepancies. For instance, if part of your USP is to create value through great customer service, but your plan shows you selling through big box retailers who don't care about your product (other than to sell it), you have a very weak story.

You must also consider timing, such as the timing of sales tactics matching your revenue timing. Say you expect rapid sales through a direct sales force. You must explain how you will ramp up a sales team to match the revenue, bearing in mind that it takes time to hire new sales people and find distribution channels and even more time for them to get up to speed on effectively selling a product.

Step 6: Describe Key Marketing Tactics

One of the key variables you had to estimate when developing your revenue forecast was a customer take rate. To effectively defend your take rate and prove that it is possible, you'll have to describe specific, potential marketing tactics. You won't have a lot of time now to craft compelling messages, develop marketing materials, build a website, or create a show stopping video guaranteed to go viral. However, you can spend time and work with creative marketers to identify a couple of key potential activities that match your value proposition and get customers excited.

I've seen many average ideas make it through the executive inquisition simply because the innovator (who usually had marketing experience) took a little time to develop a preliminary product brochure or a short, basic video that explained the concept, USP, and key attributes in a compelling, visual way. Executives are no different than consumers. The sizzle of great marketing, even if it's just a preliminary set of tactics, will help sell your idea. Just remember at this point, don't go into detail. Stick to ideas, but make them compelling, match your USP, and get executives to believe that the tactics will attract customers and convince them to purchase. If you're a technical person, developing these tactics may be a challenge and a good reason why partnering with an internal marketing person is critical.

Step 7: Describe Your Risk Mitigation Plan

Because all of these tactics are preliminary at this point, you won't have a lot of details, which means you have uncertainty; which translates to risk. It's time to consider how you will mitigate and communicate the *execution risk* as discussed in Chapter 5. For example, if you don't have the skills to develop the technology, how will you get them? If you don't have the internal marketing skills to enter a new market, how will you build them? No need to list every possible problem you'll encounter, but do understand the issues that will raise the most eyebrows. And be prepared to respond to those questions with satisfying solutions.

Breakthrough R&D might need special consideration

Facing advanced technology creates unique challenges when innovators are attempting to justify the investment. Sure, if an R&D team can solve a big problem, then there could be a big market for the new technology. Many technologists over time have developed new technology with radical benefits where nobody believed there would be a market. When the computer was invented (that is, a room full of vacuum tubes), then CEO of IBM, Thomas Watson, is reported to have said, "I think there is a world market for maybe five computers." The telephone was seen as a toy, and the TV as well was thought to be a fad. These are historic events with priceless results, but you can't underestimate the power of a company's status quo and the financial reliance on existing technology.

Even though many R&D projects are considered "experiments," R&D labs should have a basis for prioritizing projects along with every other innovation project in a company. In other words, they should all go through a similar 30-day investigation, as you're doing now, to focus the R&D effort and justify whether the project is worth moving forward. It may not be a specific P&L you're trying to develop, but at least an understanding of the customer problem and how much money is being spent on it.

In any case, whether your R&D project is fantastical, like an energy-to-mass converter a la *Star Trek*, or more down to earth, such as a Band-Aid™ that can rapidly detect cancer cells, you still have to show a clear approach to your R&D breakthrough in order to justify funding.

CHAPTER 13 REVIEW—DEVELOP A PRELIMINARY GO-TO-MARKET PLAN

At this stage of idea development, it's usually not necessary to finalize specific go-to-market tactics. However, thinking through the basic elements early on allows your executives to visualize the concept being developed, marketed, and sold. Considering these tactics now is also critical for developing a more complete story of the financials in upcoming steps. And you'll be able to highlight key areas that require further investigation once the concept is approved.

Savvy Success Timetable

Times will vary by project and executives' needs, but a savvy innovator should use a balanced approach to address every aspect of an opportunity. As a guideline, here are the estimated hours you should spend following these steps of your 30-Day Action Plan to develop a preliminary go-to-market plan:

Step 1: Create a unique selling proposition	1 hour
Step 2: Develop a preliminary product roadmap	2 hours
Step 3: Clarify your enabling capabilities	2 hours
Step 4: Develop a preliminary R&D plan	4 hours
Step 5: Describe key sales and channel tactics	2 hours
Step 6: Describe key marketing tactics	2 hours
Step 7: Describe your risk mitigation plan	1 hour
Total:	**14 hours**

MossBeGone
Mark's 30-Day Action Plan (continued)

Objective 4: Develop A Preliminary Go-To-Market Plan

Mark is ready to convince executives that he understands the tactics of taking his concept to market. As an engineer, he's used to developing R&D tactics, but putting together a preliminary go-to-market plan will be stepping out of his comfort zone.

Step 1: Create a unique selling proposition

Mark starts by developing the following USP from the customer information he gathered earlier:

For homeowners living in regions where roof contaminants are rampant, the MossBeGone roof-cleaning robot provides a safe, effective, autonomous roof cleaning solution. Compared to roof-cleaning services, the MossBeGone robot provides a labor-less and affordable alternative.

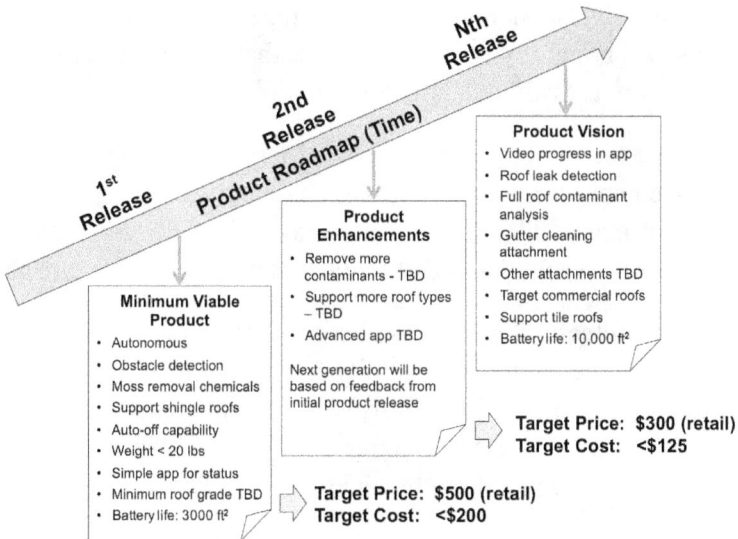

Figure 13.2: MossBeGone preliminary product roadmap

Step 2: Develop a preliminary product roadmap

For his preliminary product roadmap, Mark first articulates a longer-term vision for his concept and then adds a series of product releases that will lead to the vision. He ensures this first iteration includes the most important attributes necessary to enter the market successfully. Mark's preliminary roadmap is shown in Figure 13.2.

Step 3: Clarify your enabling capabilities

Mark has a range of technology trends to potentially leverage for creating a market advantage. Some were obvious, but others would need further research. He narrows it down to the following capabilities that would lead to a unique product with the most value for customers:

Mobile applications: These could give the user a status update of the robot's progress. An application would add both functional value as well as assist marketing efforts by creating a more exciting product that could lead to more word-of-mouth communication by consumers.

Battery technology: The robot's operating life is a major challenge. Advancements in battery storage and recharging technology will hopefully provide solutions that allow the device to operate longer without a power cord.

Sensor technology: A major trend Mark is counting on is the advancement of sensor technology to detect specific types of roof contaminants (like moss or mold) as well as edges, cleanliness, roof safety problems and other factors that customers care about.

Traction technology: For safety reasons, Mark especially needs to overcome the robot's stability factors on the roof. He projects that advancements in stability sensors and components used in automobiles and self-balancing, two-wheel transportation devices (like the Segway) might lead to a better solution.

Step 4: Develop a preliminary R&D plan

Next Mark works with his R&D team to project the skills, resources, and time it will take to develop the first version of the product. Since his team doesn't have all the necessary skills, he makes some educated guesses and identifies how to fill in the missing pieces. He also discusses the project with the manufacturing team to estimate some of the capital costs required to manufacture the robot. However, since RoboCo only produces industrial robots, Mark decides to call a contract manufacturing company that develops consumer electronics devices. After putting a non-disclosure agreement in place, they review his early prototype drawings and his rough description of the product. Based on all his research, Mark develops the following R&D preliminary headcount estimates and development timeline.

Year 1		Year 2	
Key milestones:		**Key milestones:**	
Early prototype	6 Mo.	Pre-production	14 Mo.
Beta product	12 Mo.	First product ship	18 Mo.
Management:		**Manufacturing:**	
Product manager	1	Production lead	5
Project owner	1	Manufacturing eng.	1
Device technology:		Miscellaneous TBD	1
Engineering manager	1	**Additional headcount:**	
Architect	1	Add'l R&D:	
Electrical engineers	1	TBD	3
Mechanical engineer	3	**Add'l manufacturing:**	
Application development:		Supervisor	1
Application lead	5	Manufacturing Eng.	1
App developers	2	Miscellaneous TBD	4
Total R&D Headcount:	10.5	**Total R&D Headcount:**	14
Total Manufacturing:	2.5	**Total Manufacturing:**	8.5

Figure 13.3: MossBeGone preliminary R&D estimates

Step 5: Describe key sales and channel tactics

Knowing that he needs a little help with sale tactics, Mark turns to his one sales manager with retail experience. Together they develop an overview of the key tactics and expenses necessary to take the product into home improvement retailers. He summarizes the tactics as follows:

- Primary sales efforts will be through home supply retailers. Large retailers will purchase directly through RoboCo. Smaller retailers will purchase through distributors that will be identified later.

- A new sales team will be needed that has experience selling to retailers. We'll need to work on the estimate of headcount and organizational structure, but initial projections include:

    - A business development lead with experience doing deals with big box retailers.

    - A sales leader and at least three sales professionals.

- RoboCo will also sell directly to consumers online. The website will also help customers find retailers, answer questions, and provide pre- and post-sales support.

Step 6: Describe key marketing tactics

Now Mark turns to one of the marketing managers he worked well with on a previous project to brainstorm some preliminary key marketing tactics. They are summarized as follows:

- Given the objections some customers had, we'll need to develop a series of demonstration videos to prove the robot can effectively and safely clean roofs.

- We'll do some pre-launch customer trials to capture a range of positive customer testimonials for the official product launch.

- To generate attention and create a buzz, we'll set up small demonstration displays in high-traffic areas. The logistics, costs, and effectiveness will need investigating.

- Ideally an early spring launch will boost initial sales and gain fast attention as it's the time many homeowners are thinking about roof cleaning.

Step 7: Describe your risk mitigation plan

As Mark thinks about objections during the executive inquisition, he expects they'll center on market forecasts and his ability to create demand. He knows his numbers are mere estimates at this point, and while he believes his customer data is valid, much of his forecasts are based on speculation. To lower risk, he'll need to communicate this openly. The cost of manufacturing the product is also a significant risk. He'll have to address each of these in his MRA.

Mark's next task is to develop pro forma financials to determine whether the concept will be financially viable for RoboCo.

(To be continued...)

Chapter 14

DETERMINE THE VALUE TO THE COMPANY

Put The Financial Story Together, Including Pro Forma Financials And ROI Estimates

Financial statements tell your entire story, so they must tell the same tale as the rest of your proposal.

Savvy Steps To Success

Step 1: Identify an Acceptable Pro Forma Financials Format

Step 2: Develop Preliminary Cost Estimates

Step 3: Develop Preliminary Expense Estimates

Step 4: Calculate Profit and Return on Investment

Step 5: Validate Your Calculations and Results

Introduction

With forecasts for key variables and a preliminary tactical plan as described in previous steps, you have all the information necessary to develop a pro forma P&L statement and estimate an ROI. This one page of numbers may be all executives need to see since they'll first and foremost focus on the financials of a new opportunity.

If you like spreadsheets, you may have been tempted to make a P&L statement the second you had your idea and got excited by the intoxicating profit and huge ROI. That's not a bad thing. But it's easy to make a spreadsheet say anything you want. Executives know this and they'll often assume you've manipulated market data, customer input and product development efforts to make your idea look good. So to avoid the perception that you're sandbagging, build your pro forma P&L from the results of all of the previous steps we covered so far. Don't just try to justify your pre-ordained P&L. The good news is once you've successfully completed the previous steps, doing financial statements is very straightforward.

Let's get started.

Step 1: Identify An Acceptable Pro Forma Financials Format

Communicating the complexity of your idea, market, competitors, customer needs, forecasts, tactical plans, etc. is a daunting task. And it may come down to one page—your pro forma P&L statement. This one page is the numerical culmination of every element of your story. Good executives can read your entire story just by looking at your financials, and it's not unusual to hand them a whole presentation and see them go directly to the P&L.

Once they do, they'll quickly formulate a picture that looks something like what is shown in Figure 14.1. Depending on your opportunity, it may have a finite period of profit or it may continue for the foreseeable future. In either case, from this one picture, executives can see the relative magnitude of the investment versus profit and can identify the risk associated with key metrics (such as time to profit and how long you expect to make profit). What they see will immediately tell them whether your idea has merit and whether you're an innovator they should take seriously. Assuming they find your P&L acceptable, they may probe further with these questions:

- Where did you get these revenue forecasts?
- Where did you get these cost estimates?
- Where did you get these expense estimates?

Make no mistake…if you don't have clear, concise and thoughtful responses, the rest of the discussion will not go well.

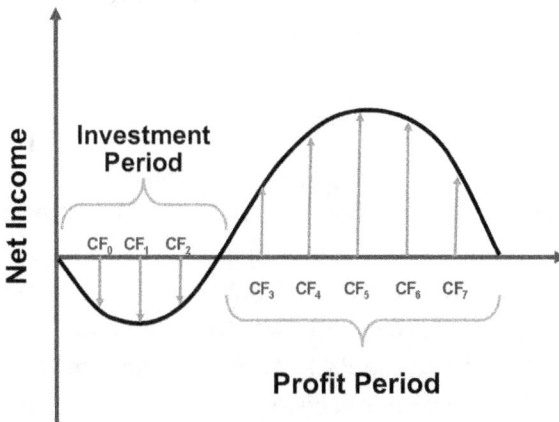

Figure 14.1: The profit and loss curve

Find the right P&L format

Your first step in the right direction is to obtain a P&L format that executives are used to seeing. You might be tempted to create your own spreadsheet and format, but I advise against this. Using a format that your executives recognize not only helps them scan it quickly and ask questions, but also gives you immediate credibility because it shows them you're not an outsider. It says that you understand the internal workings of the company and are comfortable discussing financial concepts using the same terms, line items, and calculations that they do.

It shouldn't be difficult to find a good format. You may already even have a P&L format that's appropriate for your ideas. If not, seek out a financial analyst or other finance person responsible for looking at new opportunities and ask them for a simple format based on the type of idea you're proposing. If necessary, go right to the CFO. He or she will have a variety of P&L formats to choose from. If it looks too complex (which it always does), specify that you're planning to present an opportunity for the first time and need something that includes only the most basic P&L line items. And hey, while you're at it, now's a good time to ask for support on developing your financials. You'll need this person in Step 5.

Start with your revenue estimates

Once you have a format, you can start developing your P&L statement by adding the revenue estimates you developed from Chapter 11. Because revenue forecasts are often a separate model based on estimations of market sizes, target markets, and take rates, you'll usually develop your revenue estimates on a separate worksheet and carry the results over to the P&L. However, there really is no right format other than what your decision-makers need to see.

Step 2: Develop Preliminary Cost Estimates

One number that gets a lot of scrutiny is your estimate of gross margin. Recall from our 30-Minute MBA, the major component of gross margin is Cost of Goods Sold (COGS) or Cost of Sales. Gross margin is an important financial metric for most companies since most current products are judged by their gross margins. A product that contributes 40% gross margin to profit is obviously superior to one that delivers a 20% gross margin. A higher gross margin produces more money for sales, marketing, executive bonuses and other operating expenses, or simply, more profit for the bottom line.

Estimating costs for a new product or service opportunity seems straightforward enough, but executives will always scrutinize the numbers. They're usually quite familiar with the cost structure of current products and services so they can confidently poke holes in your data. Right now you don't need a detailed bill of materials (BOM) or exact list of costs. However, whatever you estimate for COGS must be believable enough to remove any doubts that you can achieve a level of profit and margins that are acceptable to executive and corporate goals.

Most innovators are comfortable estimating the cost of parts, but other items like warranty, service, manufacturing, delivery, etc. are not so clear and might be difficult to calculate for new concepts. You'll be asked questions such as, "Have you built in our standard overhead into your COGS?" or "Have you built in the fully burdened cost of labor?" Working with your finance team can prepare you to answer with a quick yes so you can move on.

Concepts that differ from current products need more thought

Developing COGS for an idea that's drastically different from your company's current products or services creates a problem since executives don't know enough to question your numbers (but they will anyway). To get good estimates, you may need to research outside of your company. For example, if your company currently builds high-end, high-cost equipment and you want to develop a low-cost product or one targeted to a different market, you'd need to find suppliers or manufacturers that have experience in the new product to estimate believable COGS.

Step 3: Develop Preliminary Expense Estimates

Next to complete on your P&L are the line items for operating expenses. You certainly won't have full marketing or development plans, but you can use your preliminary plans from the previous chapter to get an idea of the expected expenses. The operating expenses in a P&L are typically in three buckets:

Research and development (R&D): The pre-launch and on-going expenses related to product development including personnel, equipment, and services.

Sales and marketing (S&M): The pre-launch and ongoing expenses related to selling and marketing.

General and administrative (G&A): The allocation for various overhead to cover office, management salaries, and miscellaneous operating expenses.

Note two major phases of operating expenses to consider: *pre-launch* and *post-launch*. Pre-launch expenses are related to developing and launching your concept up until it's ready for deployment. Developing your go-to-market plan in the previous chapter focused on pre-launch expenses. Post-launch expenses are the on-going expenses for your concept after launch. You'll also have to consider how to estimate these based on your company's current expense structures. For example, if current products usually get allocated an expense equal to 7% of revenue, then this is the number you would use for later years of your forecast. If your forecasted expenses are dramatically under current allocations, then this raises a flag that your expenses are too low. The same allocation method would hold for R&D and S&M expenses.

However, like with COGS, if your concept doesn't look like your company's existing products, there is danger in applying the same expense structure. A savvy innovator understands the expense structure of a company's current products and can defend a different structure when necessary. Smart executives realize that some innovation efforts require a new business unit. They might allow a team to operate under different expectations for expenses and related margins than existing products. For example, IBM had to develop their IBM PC project under a different business unit. Likewise, the *New York Times* digital edition had to operate separately from the newspaper business unit.

Don't forget to think like an entrepreneur

Before accepting internal estimates for anything, be it R&D expenses to the cost of creating a brochure, think about what the expenses might be if you hired services outside of your company or used different vendors. Large companies use large vendors and pay top dollar for everything from consultants to logo development. But for really new opportunities, you can lower risk by using lower-cost vendors and finding cheaper alternatives.

In the last chapter, we read about Paul, who asked his internal team to develop R&D estimates for his new social media experience around sharing video. Because of the team's high expense structure and the unknown timeframe it would take to develop the application, they came back with an outrageous estimate of $3.7 million to develop the first version. Knowing this amount would not be acceptable, Paul went to outside vendors who gave bids ranging from less than $750,000 to $2.5 million. The risk of a $750,000 project is obviously much less than a $3.7 million project. When preparing to face the executive inquisition, he included in his MRA a milestone for vendor selection and approval. Or alternatively, help to work more effectively with the internal R&D.

However, you must keep in mind that using outside vendors for new opportunities is one of the major political challenges innovators face. Going outside for key services almost immediate triggers the corporate immunity system into action. So be prepared.

Step 4: Calculate Profit And Return On Investment

Once you have appropriate revenue, costs and expenses for each period of your pro forma P&L, calculating profit and ROI is pretty straightforward—almost. Because these numbers are so critical to a funding decision, you must be sure to calculate them in the ways your decision-makers want to see.

There are different ways to calculate profit and ROI. They're generic finance terms so every company will have its own way of calculating them. Some may even use different terms, like EBIT (Earnings

---

Before Income Taxes), or EBITDA (Earnings Before Income Taxes, Depreciation and Amortization), or simply 'operating profit' among others.

For ROI calculations, most companies use some form of Net Present Value (NPV) or Internal Rate of Return (IRR), but even these can have variations. For example, did you know that Microsoft Excel assumes that each cash flow occurs at the end of the period and therefore calculates an NPV accordingly? But if you calculate NPV based on the cash outflow (or inflow) occurring at the beginning of the period, you'll get a different result. Also the treatment of risk in ROI calculations varies from company to company. Some prefer to calculate ROI using a higher hurdle rate for high-risk opportunities. Others want to see various scenarios of revenue and expenses to take risk into account.

The point is a savvy innovator develops financials using the exact information and calculations that executives need to see. If you assume you already know all of this and don't take the time to use an accepted P&L template and get help from finance colleagues, be prepared for serious opposition to your numbers. Questions about definitions, calculations, and determinations of key variables are very distracting from the real issues related to your idea.

Step 5: Validate Your Calculations And Results

We've already discussed in depth the importance of getting feedback on your forecasts and financials, but do not skip this last step. Have an ally, a respected finance person, who can review and validate your financials and back you up in front of executives. Seek your allies far in advance of the executive inquisition because:

- They can provide a template with the right calculations, terminology, and expected line items.
- They can guide you to the information you'll need for revenue, COGS and expenses.
- They can help you determine how to best treat capital expenses, depreciation, allocations, and other financial elements.

- They can validate your calculations of profits, margins, and ROI.

- They can guide you on how to discuss and account for the risk in your financials.

Inevitably a decision-maker will ask you, "Have you worked with the finance team on this?" How great to answer with a resounding, "Yes! They helped develop the P&L and are very supportive of this plan!" But here's the catch…the question wasn't really directed at you. She was looking at the senior financial person in the room to see if that person answers with a similar, but perhaps less enthusiastic, response.

CHAPTER 14 REVIEW—DETERMINE THE VALUE TO THE COMPANY

Creating a P&L statement is rarely a problem. Creating one that will be accepted by decision-makers is a far more daunting task—one that must be completed with your finance team. Each line item must be developed using reasonable data and expert input. Even at this early stage, your numbers will be scrutinized to ensure your opportunity has the potential to meet the financial targets of the company. You need to at least get acceptance of the magnitude and direction of your revenue, costs, expenses, and every other aspect of your P&L before you'll get acceptance of the potential profit or ROI.

Savvy Success Timetable

Times will vary by project and executives' needs, but a savvy innovator should use a balanced approach to address every aspect of an opportunity. As a guideline, here are the estimated hours you should spend following these steps of your 30-Day Action Plan to determine the value to the company:

Step 1: Identify an acceptable pro forma financials format	1 hour
Step 2: Develop preliminary cost estimates	2 hours
Step 3: Develop preliminary expense estimates	2 hours
Step 4: Calculate profit and return on investment	2 hours
Step 5: Validate your calculations and results	3 hours
Total:	**10 hours**

MossBeGone
Mark's 30-Day Action Plan (continued)

Objective 5: Determine The Value To The Company

Mark has all the information he needs now to develop a full pro forma P&L statement. While much is still unknown about his concept, he must prove to executives that RoboCo margin and profit targets are achievable and that he has a path to reaching them.

Step 1: Identify an acceptable pro forma financials format

When Mark spoke with Tony about the revenue forecasts, he also asked him for a template to use for a financial P&L. Tony sends Mark to see Betty, one of his financial managers, who was responsible for developing forecasts for RoboCo's current products. She gives him a format that he modifies as shown in Figure 14.2.

Step 2: Develop preliminary cost estimates

To complete the COGS, Mark has to get resourceful because his internal information is limited since RoboCo has never built a consumer robot. He knows the company wants to see gross margins of >50%, but he's hoping (a dangerous proposition) they'll accept margins of 35% for a consumer device.

Using his short-term wholesale price target of $325 and the longer-term target of $195, Mark calculates cost targets of approximately $200 at launch that should quickly get cost reduced to less than $125. He needs to determine if this will be close to achievable. Unfortunately, RoboCo's supplier and production team is of little help, so again, Mark calls the contract manufacturing company that develops consumer electronics devices he talked with previously. They estimate the product could be built between $180 and $325 per unit, but over time, the cost could come down substantially. The variability is wide because of the unknown volumes, complexity of production, and selection of components. To allow some room for error, Mark estimates a starting cost of $230 that comes down to his target of of $125 in year five. The contract manufacturer also helps Mark estimate the capital costs and steps required to get a product like the MossBeGone into manufacturing. Now Mark thinks he has enough information to hold an intelligent conversation, but he needs more guidance on the best way to present it at the executive inquisition.

Step 3: Develop preliminary expense estimates

Although Mark has some preliminary R&D, sales, and marketing tactics, he now has to estimate the total expense of employees. Mark visits Betty again to develop expense estimates for the expected headcount. She suggests Mark use an average fully-burdened expense of $180,000/employee. This will obviously be different for managers versus non-managers and for various roles, but it was close enough for a pro forma P&L that could be refined later.

Mark also estimates overall expenses for sales and marketing costs to launch the product based on his discussions with the marketing department as shown in the P&L statement (Figure 14.2). He doesn't bother breaking out headcount from other marketing expenses.

Step 4: Calculate profit and return on investment

Calculations for profit and ROI turn out to be straightforward. Betty provides the following guidance as well:

- Use operating profit as his bottom line and don't worry about taxes or interest.
- Ignore depreciating or amortizing capital expenses for now (they will revise these calculations later if the project moves forward).
- Calculate a hurdle rate of 13% in his NPV.
- Use allocations of 10% for R&D expenses and 8% for sales and marketing expenses after year two, as well as 6% for general and administrative expenses for all years.

With all the numbers in place, Mark looks at his P&L statement as shown in Figure 14.2. "Hmm," he thinks, "would I invest in this?" The numbers didn't look as good as he'd hoped. He believes the revenue looks enticing, but when he takes into account all of the R&D investment and costs, the NPV is only $11.7 million on an investment of over $11 million before hitting cash flow positive in year three.

He wonders how to proceed. Should he manipulate the numbers to make the investment look better? "If I just double the take rate, this will look really good," he thinks. He decides to keep the pro forma calculations as they are for now and see what happens in his next steps.

Step 5: Validate your calculations and results

Mark works closely with Betty to refine the numbers, as well as the allocations to use for each type of expense over time. She can't provide feedback on the validity of the forecasts, but she does validate that the P&L provides the right type of information. She assures Mark has the right calculations that would meet Tony's needs when it came time to review the project's financial potential. However, Mark wonders what his chances are of gaining support for his numbers in advance of the inquisition.

He asks Betty, "What do you really think of this concept as an investment?"

She thinks a minute, "You've got a battle on your hands. This will be a lot of risk for Tony to accept, but if Grant wants to get into the consumer robotics market, he might take the risk. I'm looking forward to the discussion myself."

Mark replies, "Do you think I should develop best and worst case scenarios to show the potential if it was a huge success?"

"I don't think so. The numbers speak for themselves. This is good enough for a discussion. If they're interested, there'll be many more rounds of financial analysis. We can work up more scenarios then."

"Ok...Sounds good."

With his financial information ready, Mark thinks ahead to the final format and presentation for his proposal. He'll only have a short amount of time to get his executives excited about MossBeGone so clearly he has some more work to do.

(To be continued…)

Moss-Be-Gone 7-Year Pro forma Profit and Loss

	Investment Period		Profit Period				
	Year 1	Year 2	Year 3	Year 4	Year 5	Year 6	Year 7
Revenue							
Est. Retail Price		$500	$400	$350	$300	$300	$300
Wholesale Price (Retail margin ~35%)		$325	$260	$228	$195	$195	$195
Units		13,500	118,703	316,540	411,502	506,464	506,464
Total Revenue	$0	$4,387,338	$30,862,650	$72,012,850	$80,242,890	$98,760,480	$98,760,480
COGS - Targets							
Variable		$150.00	$120.00	$90.00	$75.00	$75.00	$75.00
Fixed/unit		$80.00	$70.00	$60.00	$50.00	$50.00	$50.00
COGS/unit	n/a	$230.00	$190.00	$150.00	$125.00	$125.00	$125.00
Total COGS		$3,104,885	$22,553,475	$47,481,000	$51,437,750	$63,308,000	$63,308,000
Gross Income		$1,282,453	$8,309,175	$24,531,850	$28,805,140	$35,452,480	$35,452,480
Gross Margin/unit		$95.00	$70.00	$77.50	$70.00	$70.00	$70.00
Gross Margin (%)		29%	27%	34%	36%	36%	36%
Expenses							
Total Headcount	13	20					
R&D Headcount expenses ($180K)	$2,340,000	$3,600,000					
Capital expenses	$1,500,000	$3,000,000					
Total R&D (10% after year 2)	$3,840,000	$6,600,000	$3,086,265	$7,201,285	$8,024,289	$9,876,048	$9,876,048
S&M (8% after year 2)	$250,000	$1,500,000	$2,469,012	$5,761,028	$6,419,431	$7,900,838	$7,900,838
G&A (6%)		$263,240	$1,851,759	$4,320,771	$4,814,573	$5,925,629	$5,925,629
Total Opex	$4,090,000	$8,363,240	$7,407,036	$17,283,084	$19,258,294	$23,702,515	$23,702,515
Operating Income	$(4,090,000)	$(7,080,788)	$902,139	$7,248,766	$9,546,846	$11,749,965	$11,749,965
Op.Income/unit		$(524.52)	$7.60	$22.90	$23.20	$23.20	$23.20
Operating Margin	n/a	n/a	3%	10%	12%	12%	12%
ROI Calculations							
NPV (13%)		$11,726,086					
IRR		41%					

Figure 14.2 MossBeGone 7-year pro formal P&L

Chapter 15

MAKE THE PROPOSAL CONCISE, COMPELLING AND COMPLETE

Fill In Missing Gaps To Create A Succinct Proposal That Includes Your MRA

If you can't get your idea across in ten presentation slides, adding another 30 won't help.

Savvy Steps To Success

Step 1: Determine How to Present Your Story

Step 2: Simplify Complex Information

Step 3: Develop Your Opportunity Scorecard

Step 4: Develop Your MRA

Step 5: Close Gaps, Cut Garbage and Remove Confusion

Introduction

At this point, your opportunity proposal is taking shape. You have a great concept, input from real potential customers, acceptable forecasts, and a clear pro forma P&L statement. You might also have pages upon pages of notes and dozens of presentation slides. This is all good, but also creates your biggest problem—you are TOO knowledgeable. When you face the executive inquisition, it'll be easy to forget that you're way ahead of your audience in knowledge, conclusions, and conviction. You've devoted a lot of time and sweat into your idea now—you may have also lost objectivity.

This chapter focuses on the three C's of a great opportunity proposal: Concise, Compelling, and Complete. It must tell a story in a way that is easily digestible to your executive audience and makes them want to give you the resources to take your idea further. The time you spend now to simplify and clarify your message can make the difference between a powerful proposal that leads to a "Yes!" or a mediocre proposal that is as confusing as all the other ones that might be heard that day.

Let's get started.

Step 1: Determine How To Present Your Story

In the first step of the 30-Day Action Plan, you were to find the template your company uses for opportunity proposals. This was important for identifying the sections and type of information to include in your proposal. However, you won't actually be presenting that template as your proposal. Why? Because the structure and level of information is designed to conform to corporate needs rather than tell a compelling story.

Your final proposal should serve two unique purposes, each requiring a different type of document.

1. **To explain your opportunity to someone who will read about it.** This document must convince the lone reader that your opportunity is worth funding, and you must address key objections without the benefit of further explanation.

2. **To inspire a live audience.** This document must quickly win over a room full of skeptics, quickly corral disparate opinions, and focus everyone's attention on agreeing to fund your idea.

Your documents must be concise, compelling, and complete. But if you try to solve both purposes with only one document, like a lengthy and detailed PowerPoint presentation, you'll rarely fulfill both purposes.

Identify the right document format

I see A LOT of presentation slides. That's not a bad thing, but not always the desired format. Amazon's CEO, Jeff Bezos, has become infamous for declaring he doesn't want to see another PowerPoint presentation. He prefers short documents with concise paragraphs, graphics and corresponding data. During review meetings, he allows a period of time for everyone to read the document before they discuss the poignant issues. He knows most people won't read the proposal in advance, so this solves the problem of half of the people not being as prepared as others. I personally like this approach since it allows everyone to get a complete picture at the same time with the right level of detail before discussing.

Ideally you should ask what decision-makers want to see; both for length and format. I remember one team getting ready to share their proposal for a new mobile application. They asked the CEO how he'd like to see the proposal. He replied, "I want a nine-page document." Not eight. Not ten. But exactly nine. This was based on his experience that nine pages was the right length to tell a story succinctly with enough detail to be complete. It seems that using a short document instead of presentation slides is becoming a trend.

Our focus: The Executive Inquisition

Some people will be reading your opportunity proposal from a document, but in almost all situations, you'll be sharing your idea in front of a live executive audience in the form of a presentation. This meeting often ends up feeling like an inquisition. Not only will your

---

idea be placed under a microscope, but you'll need to defend your numbers, your plans, and possibly even your reputation. The clearer and more powerfully you can present your proposal, the greater your chance for success. While much of the upcoming tactics are important for any form of proposal, they are critical if you want to succeed during the executive inquisition.

Telling stories

Your executive audience will have a range of mindsets, from analytical to emotional, that will respond to different types of messages, data and visuals. You have to appeal to all of them. One technique that's universally appealing is the use of story-telling. No matter where I travel, I'm always amazed at the power of stories. I can be in China talking about an important marketing technique or a fascinating industry example, and as soon as I utter the words, "Let me share a story from when I was…," I see an immediate change in my audience. The room gets silent, eyes open wide, and people lean forward. Everyone likes a story and smart leaders from Abraham Lincoln to Steve Jobs have used the art of storytelling to sell an idea. When telling a story to sell your idea, the main character, however, is not your idea, but rather your target customer. The plot is how you will win in the market. The exciting action scenes are how you validated your concept. And the happy ending is the potential profit. By helping the audience visualize a real customer solving a real problem and getting real results, they will quickly see why they should pay attention to the rest of your proposal.

If you've ever participated in Toastmaster's International, you might be familiar with Monroe's Motivational Sequence, a presentation flow developed in the 1930's by Alan Monroe at Purdue University. It's designed to persuade an audience to move forward and accept your recommendation. Following are the steps for this sequence, which I've modified slightly to meet the needs of innovators.

Attention. Get the attention of your audience by using a customer story, interesting example or dramatic statistic related to the problem your idea is solving.

Need. Demonstrate how big the need or problem is for customers and help the audience visualize the gap in the market for filling this need. This step should also show how important the opportunity is for your company.

Satisfaction. Demonstrate how your idea solves the customer problem. Convince your audience that your concept solves the need, has real value, and entices customers to take action.

Visualization. Help decision-makers visualize how you will execute key tactics to move your idea forward and fulfill a vision. They must to be able to see how it will unfold with specific actions ranging from developing technology, strategizing marketing activities and acquiring customers.

Action. Describe the specific short-term, call-to-action necessary to move your opportunity forward with minimum risk—the MRA. Specific steps and milestones that further validate your concept, lower risk, and prove you can execute will have the most power to influence executives.

When done properly, this presentation flow technique will allow you to share a range of visuals, data and analytical information without getting bogged down by too much. You'll stay focused on the goal of selling your idea and getting funding.

Step 2: Simplify Complex Information

In most meetings, people start tuning out after about 20 minutes. If you haven't captured their attention in the first 2-3 minutes, they're mentally off to email, their next meeting, or what they're having for lunch. While it's taken you 30 days to prepare your proposal and develop your conclusions, you need your audience to come to the same conclusions in a fraction of this time. It's important to hit the highlights early and quickly demonstrate value with key supporting information. Grab them in the first three minutes and convince them that what you're selling is valuable enough to listen to for the next 17 minutes. If you have an hour scheduled, you can then enjoy the last 40 minutes discussing the critical next steps instead of trying to defend your data or responding to questions that should never have been asked in the first place. Do not make the mistake of developing a presentation that drags on for 60 minutes before you get to the punch line. No one has the patience or attention span for that.

A savvy innovator knows how to simplify data, analysis, and results to make even the most complex topics simple, clear, and relevant. We've all seen presentations from very smart people go bad when the presenter dives into details that the audience is not prepared for. Once this happens, it's easy to think the audience just doesn't get it, but usually it's because the innovator has not explained things in the context of the audience's current knowledge and level of understanding.

"RICH DUMMY TERMS"

You might remember the movie, *The Nutty Professor,* starring Eddie Murphy as Professor Clump and his alter ego Buddy Love. There's a scene in the movie that brilliantly makes executives look dumber than they really are. Buddy is asked by his school's dean to dumb down his scientific explanation of his fat-reducing miracle tonic. He retorts, "Oh, you mean rich dummy terms? I'll break it down for all the rich dummies in the room. Listen up." Buddy then changes his message from explaining the science behind his tonic to explaining the customer benefits. "There's a gene in your D.N.A. that routes fat straight to your fat cells, and it causes unsightly conditions. It's quite nasty. There's a way we can turn these genes off. I'm not talking about using exercise or diet. I'm talking about taking a simple solution that helps reconstruct your metabolic cellular strands, thus giving you the appearance of, in medical terms, 'gluteus minimus'."

Of course this is meant to be flippant and entertaining, but the message is correct—when you're addressing an audience that may not have the same background or technical knowledge you have, you must tune the information for them in a way they'll understand.

The one-minute explanation rule

While it's often the technical message that needs to be simplified, it can just as often be the product plan, market forecasts, or competitive landscape. I've been guilty of developing overly complex models that try to show customer needs or market segments then expect my audience to be wowed and draw the same conclusions that I did. They rarely do. One quick way to determine if a concept is too complicated is to test how long a slide takes to explain. If someone can't grasp the information in less than one minute, it's too complicated.

Leave overused tools in your toolbox

Many innovators are using common business analysis models in their opportunity proposals. Popular ones include: Porter's Five Forces, the Boston Consulting Group's Market Growth-Market Share model (where the terms 'cash cow' and 'dogs' come from), and the infamous SWOT model (Strengths, Weaknesses, Opportunities, and Threats). But I urge you to think twice about including them. Take SWOT for instance, the one I see overused and abused most often. The innovator has a one-slide SWOT analysis with high-level bullets in each category (like 'Broad Distribution Channels' listed as a "strength"). I always wonder, "So what?" When I later ask the innovator why the SWOT analysis was included, I usually hear, "Because I thought I needed it."

The truth is, most executives know all these factors and will automatically build them into their internal perceptions of your idea. These models can be valuable for explaining strategic thinking, but unless you really need them to provide specific rationale for your idea, leave them out (or stick them in a back up slide). Besides, they just take time to explain, and executives usually prefer that you stick to the facts and tell them something they didn't know.

The dos and don'ts of presenting data

Because you're the expert on your concept and know more than anyone else about it, it's tempting to share everything you know and have discovered. But people can only assimilate so much information at once. It's critical to use data during your opportunity proposal, but you must curate and communicate it in the most simple and powerful way possible. The key? Concise messages, clear graphics and simple models that tell your story and no more. Save detailed data and complex models for back up. When you're preparing your data for presentation, follow these Dos and Don'ts:

Data Category	Do	Don't
Customer Value Proposition	Provide a prioritized list of customer needs and how your concept fulfills important needs.	Provide a list of features and bullets of how great your idea is.
Financial statements	Present the most important financial information—revenue, costs, expenses and margins using graphs where possible instead of detailed tables.	Present a complex spread sheet.
Competitive analysis	Show an industry map of key competitors, how they are positioned in the market and a visual of relevant competing products.	Show a detailed list of competitors with detailed feature comparisons.
Market forecasts	Provide a clear graphic of the industry size, how you've segmented the market along with the size and definition of your target market segment.	Show a bunch of miscellaneous industry data and then provide a eureka guesstimate of a revenue forecast.
Technical diagrams	Provide a simple technical explanation and how the technology leads to unique customer benefits.	Provide detailed schematics or flow charts that take hours to explain.

Data Category	Do	Don't
Tactical plans	Provide major milestones and the key activities that show progress, reduce risk and add credibility.	Provide detailed Gantt charts and project plans.

In all cases, keep bullets and text to a minimum. If you need to use anything less than 16pt font to explain something or have more than 100 words on a slide, it's time to think about simplification. And always opt to use graphics to tell the story rather than words whenever possible.

Step 3: Develop Your Opportunity Scorecard

Now it's time to consider whether your opportunity is actually worthy of funding. Never forget that executives are constantly barraged with ideas from their peers, customers, partners, competitor announcements, news, and employees, so it's critical you meet the criteria in their head of what determines an idea worth pursuing. Depending on your company's environment, the criteria may be formally established or just personal ones to each executive. Either way, if you want to sell your idea, you need to know what these criteria are.

Criteria for a "worthy" opportunity

One tool many companies use to determine an idea's "worthiness" is an *opportunity scorecard*. Much like the Five-Point Inspection we covered earlier that determines whether your idea is worthy of developing a proposal, an opportunity scorecard includes a set of decision criteria and a rating scale for each criterion related to investing. Even though many criteria are the same, how each company defines the criteria will be different. Never assume you know the right criteria for your situation. Following are some typical criteria companies use to determine if an opportunity should move forward:

- **Does the concept clearly fit with the company's strategy and direction?** While this criterion is critical, it's often the most difficult to rate since the answer is subject to the mind of funding executives.

- **Does the concept have a large enough potential?** This is usually stated as a hurdle to exclude opportunities that are not big enough to justify time or resources. For example, if your company had $3 billion in revenue last year, executives may not want to see any concepts that can't achieve at least $25 million within three years.

- **Does the concept have acceptable risk?** This one should lead to a good discussion on what risk really means. This is often a combination of commercial or technical risk, or they may be two separate criteria. As discussed before, risk is in the eye of the decision-makers, so you should rate risk based on their perception, not yours.

- **Does the concept leverage your company's competencies?** Fitting with competencies is often a criterion, but it has a way of thwarting innovation and forcing new opportunities to simply look like what the company is currently doing. With innovation you may have to build new competencies.

- **Does the concept meet other financial metrics?** In addition to revenue goals, most companies have a set of financial metrics they need new opportunities to meet. These might be gross margin targets of >40% or a maximum R&D investment of $10 million, for example. Or you might hear, "The opportunity should provide positive cash flow in less than two years."

You can see the inherent danger in these criteria and why many innovators exclaim, "Our management doesn't get it!". How is it possible to know whether a concept will achieve $50 million revenue and be cash flow positive within three years with a margin of greater than 40% before you've even developed the product and conducted extensive analysis?

Try to remember these criteria aren't meant to be black and white. The score by itself doesn't mean a lot unless it's compared with other opportunities. It simply helps facilitate a meaningful discussion to determine if valuable resources will be wasted on ideas that have little fit with where the company wants to go. A savvy innovator knows an executive needs to meet these criteria early in the decision process and may have to kill some great ideas to allow others through.

At this point you won't have all the data to provide 100% confidence in meeting any of these criteria. But you should have enough information to hold an intelligent discussion about whether there's a solid path to meet them.

Step 4: Develop Your MRA

Remember the MRA from Chapter 6? To recap, your MRA is a strategy to ask for a minimal amount of resources to fund your research and take your concept to the next level. Taking your idea beyond the first 30 days (when so far has mostly been your blood and sweat), requires real corporate resources of other people's time, money and expertise to gather more data, develop prototypes, explore technical solutions, define the product, etc. You're making a serious request, and an MRA is the surest way to success.

You should include two elements in your MRA:

- A small set of clear and relevant milestones.
- The specific resources you need to achieve the next milestone.

Your MRA should reduce risk and give executives confidence that any resources approved will move your idea forward and not be wasted. It is also a promise to executives that you will have achieved the milestone by the next time they review your progress.

Examples of milestones include:

- Creating a simple prototype ready to further validate the technology and/or customer value proposition.
- Conducting additional research such as interviewing a number of customers or using an industry consultant to validate revenue forecasts.
- Devising a development plan with more accurate resource estimates and product definition.

- Identifying potential partners, acquisition targets, or key customers.

- Developing an operations plan or more detailed cost-of-goods-sold evaluation.

Your MRA should be more than just a laundry list of next steps. It should be centered on focused milestones that address the key risks concerning executives so they can justify approving the first round of funding. Your request to achieve these milestones should also focus on more than just money. It should ask for any resource necessary to execute your idea. Some of those resources might include:

- **More** of your time—for everything.

- **More** of your colleagues' time—expertise on markets, customers, technology, legal, finance, etc.

- **More** of your management's time—feedback, support, advice, approval, etc.

- **Access** to technology such as R&D or systems support.

- **And of course**—real investment ($)—money for research, prototypes, data, software development, services, suppliers, patents, etc.

In most organizations there are very few resources that aren't already spoken for, and the competition is fierce. But every company wants to grow and is always looking for new opportunities and new leaders to help it grow. A savvy innovator knows the right strategies and can build a logical argument towards growth and profit that encourages decision-makers to make those resources available.

Step 5: Close Gaps, Cut Garbage, And Remove Confusion

Once you have a presentation and an opportunity document, take the time to test, edit, and improve. Walk through this 10-point checklist to see if you're presentation is ready for the next step—selling your idea.

✓ Does it demonstrate real customer value?

✓ Can you point to real customers that confirm your value proposition?

✓ Do you have market clarity and data to validate the opportunity?

✓ Does it communicate clearly the current and emerging competitive landscape?

✓ Have you clarified what will be really unique about the opportunity?

✓ Can you justify all forecasts using methods that decision-makers can support?

✓ Has the finance team validated your pro forma P&L statements as acceptable?

✓ Do you clearly explain and address both technical and market risk?

✓ Is there a clear path to execute the opportunity?

✓ Do you have an MRA showing your next steps and milestones?

CHAPTER 15 REVIEW—MAKE THE PROPOSAL CONCISE, COMPELLING, AND COMPLETE

No matter what your company, no matter what your environment, executives informally or formally review dozens to hundreds of opportunities. Within minutes (or seconds even) executives will immediately ascertain whether your pitch is worthy of their time. You're up against a lot of competing forces, so cutting through the noise requires a concise, compelling, and complete proposal that quickly gets attention and interest.

Savvy Success Timetable

Times will vary by project and executives' needs, but a savvy innovator should use a balanced approach to address every aspect of an opportunity. As a guideline, here are the estimated hours you should spend following these steps of your 30-Day Action Plan to make your proposal concise, compelling, and complete:

Step 1: Determine how to present your story	2 hours
Step 2: Simplify complex information	4 hours
Step 3: Develop your opportunity scorecard	1 hour
Step 4: Develop your MRA	1 hour
Step 5: Close gaps, cut garbage, remove confusion	2 hours
Total:	**10 hours**

MossBeGone
Mark's 30-Day Action Plan (continued)

Objective 6: Make The Proposal Concise, Compelling, And Complete

At this point, Mark has far more information than he can communicate in the short amount of time he'll have to present. He must now curate and summarize his information in a way that both quickly explains the concept and its potential in the market and inspires the executives to provide resources to move it forward.

Step 1: Determine how to present your story

As typical of an innovation project, Mark has hundreds of pages on the industry, competitors, and various elements of his concept. But he only has about 20 minutes to tell a persuasive story using 10-12 slides. He must plan carefully not to go over that time and to put any backup information at the end of his slides. However, given what he knows about his environment, he makes certain that the backup information can tell the complete story even if Mark wasn't there to present his concept.

Step 2: Simplify complex information

For each major section of the opportunity proposal, Mark condenses the information down to a relevant and impactful summary. He uses bullets sparingly and relies heavily on graphics and diagrams for his final presentation.

Step 3: Develop your opportunity scorecard

Working with his manager, Linda, Mark develops a score card using a set of five criteria. Each criterion is scored on a one to five scale (five being excellent; one being poor). He knows their ratings will be questioned, but with Linda's support, he thinks they're close enough for a healthy discussion about the opportunity. Mark's opportunity scorecard is as follows:

Fit with RoboCo's strategy: *Score 4* Mark and Linda agree that consumer robotics could be a nice growth area for RoboCo, but it's not an obvious fit with the company's current direction. They know this score will be heavily debated.

Revenue potential of >$25 million/year: *Score 4* Mark believes his revenue forecasts are accurate enough to prove that this is achievable with reasonable risk.

Time to profit of <2 Years: *Score 2* This is a difficult criteria given the estimated time to develop, market, and sell the product. Mark knows he'll need to address this problem in conversations before the executive inquisition.

Technical risk: *Score 4* There is some risk in new sensor technology and the mechanisms for cleaning a roof, but there's a reasonable path to working through any technical issues.

Commercial risk: *Score 3* RoboCo has never sold a consumer product before, so this will certainly be a challenge. The combination of the company's inexperience and the uncertainty that MossBeGone will be readily accepted by the market means that commercial risk is significant. Mark knows that other executives are likely to rate this criterion lower than a three.

With an overall score of 17 out of a possible 25, Mark believes this is good enough to move forward. However, he doesn't know if executives will agree nor how well it ranks against other opportunities. These are more gaps Mark needs to investigate before the executive inquisition.

Step 4: Develop your MRA

Mark understands his best chance for success is requesting a minimal amount of funds to get the ball rolling for MossBeGone. But what should his MRA be? He estimates the whole project will take up to $12 million before achieving positive cash flow, but he has to identify the specific milestones that will be most relevant to moving forward. After careful consideration factoring all the objections that have been raised, Mark develops an MRA of $250,000 based on the following requests:

- $150,000 to develop a prototype. Mark knows a local machine shop that's anxious to work on this project and is willing to develop a rough prototype at a reasonable cost in three months.

- $50,000 to develop a prototype of the application that would interface with the device.

- $50,000 to conduct additional market research to include another round of interviews followed by four focus groups. This includes using a consultant who can help develop more accurate market forecasts.

Mark also needs some time from key people like their star architects and well-respected product manager.

Step 5: Close gaps, cut garbage and remove confusion

To ensure his presentation will be concise, compelling and complete, he uses the 10-point checklist as discussed in Step 5 of Chapter 15. Ultimately, Mark finalizes his presentation. However, he can't be completely sure these will achieve his goals until he starts to pre-sell his opportunity to influencers.

Mark's headed down the home stretch but must secure support for his opportunity proposal before he has to face the executive inquisition.

(To be continued...)

Chapter 16

SELL THE PROPOSAL

Develop The Specific Tactics For Selling Your Idea To Company Executives

---

If you can't sell your idea before the executive inquisition, don't expect to sell it during the inquisition.

Savvy Steps to Success

Step 1: Think and Act Like a Professional Sales Person

Step 2: Pre-Sell Your Opportunity

Step 3: Prepare for the Big Day

Step 4: It's Go Time!

Step 5: Manage the Aftermath

Introduction

Now that you have a concise, compelling, and complete opportunity proposal, you're almost ready for the executive inquisition. However, selling your BIG idea requires a great deal more than what you can accomplish in a single presentation. A savvy innovator takes a systematic approach to gain support in the days leading up to the presentation—which is really when the selling process begins.

Let's get started.

Step 1: Think And Act Like A Professional Sales Person

If you've ever worked with professional sales people, you know they live and die by influencing a sale even when a customer is not immediately interested in buying. Think of your executives as those customers and yourself as the sales person who must influence them. Executives are rarely pre-disposed to buy your ideas—that's the nature of innovation. So you must learn how to sell, and your first step…is to think and act like a professional sales person.

Sales Process	Sales Mentality
Qualify the customer. Determine if the customer has a budget, authority and a timeframe for making a decision.	**Know the environment.** Navigate the entire environment around the decision maker.
Assess needs. Listen to the customer and ask probing questions to understand problems and needs.	**Think people first.** First and foremost consider the needs of everyone involved in the decision.
Propose a solution. Present your solution focusing on the needs you've uncovered.	
Remove objections. Continue selling until all objections are met and the customer is ready to buy.	**Be persistent.** Don't give up unless the buyer has made it perfectly clear that "no" is the final answer.
Negotiate and close. Close the deal by looking for a win-win situation.	**Impeccable follow up and communication.** Provide consistent, high quality responses that meet the needs of the customer.

Figure 16.1: Savvy innovators think like sales professionals

The chart in Figure 16.1 shows two sides of what makes a sales person successful. On the left is a typical sales process. Steps that start with qualifying customers and ends with closing the sale. These steps should look familiar. Your 30-Day Action Plan is essentially a sales process, and you've been following these steps. The final review meeting (in our case, the executive inquisition) is only one step in this process but it hopefully closes the sale of your idea. The right side of the diagram is where we are focused now—sales mentality. At this point, you must consider the following factors to think like a top sales person:

Know the selling environment

Table 16.2 summarizes the four opportunity environments discussed in Chapter 2. Selling into each environment requires a unique approach. Some environments, primarily Opportunity Engines, have a clear budget with decision-makers who approve funds for new opportunities. While not entirely easy, these environments usually provide the best, and clearest paths to getting your idea heard. Other environments are more difficult to manage since ideas require more selling, more data, and more influence from more risk-averse executives. In an environment where there is no real path to get in front of decision-makers, such as Tornados and Clogs, you can feel overwhelmed, particularly if you don't have direct access to influential executives. You can still overcome this challenge by taking a systematic approach and thinking like a sales professional.

Think people first

Great sales people understand that every person has critical needs that must be met before they will ever consider purchasing. Top sales people also realize that many people influence a purchase decision, but they never know exactly who or by how much. Therefore, they're respectful of everyone in the company from assistants to top executives. In other words, they know not to burn bridges because they never know who might be doing the buying. This 'think people first' rule is the foundation of influence and provides the basis of most long-term, mutually beneficial relationships. Savvy innovators know this and always take the needs of all stakeholders into account at every step of the idea selling process.

Be persistent

Moving ideas inside a company often requires consistent pressure over a long period of time. The power of persistence is well known. You can see this by watching children. My seven year-old daughter is a master! If she wants something, she starts with the direct request. Then tries to negotiate. Then pulls her cutest look. Then moves to a mild fit. And if all else fails, she brings out the extreme tantrum. Very effective. Somehow we adults seem to lose our drive for persistence. If we ask for something and don't get what we want, we move on. But consider this…you have a great idea and send one email or hold one meeting that gets a negative response. If you never ask again, here is what an executive hears, "I have an idea, but it's not very important, and, uh, I don't believe in it too much." If you don't get support on the first attempt, I'm not suggesting you throw a tantrum but you do need to try some other approaches until you've exhausted all angles.

		Funding Executives	Key Influencers	Budget	Decision Forums	Selling Focus
High Process	Opportunity Engine	Innovation Committee Leader	Idea sponsors	Dedicated innovation budget	Opportunity review meetings	Collaborate with executive sponsors.
	Opportunity Siphon	CEO/CFO	Sales and Technical experts	Part of operating plan	Operations/ product reviews	Manage the process.
Low Process	Opportunity Tornado	Varies – One per idea	Many – Anyone with power and budget	Slush funds, pet projects	Hallways & private discussions	Sell to influential executives.
	Opportunity Clog	CEO/GM	CFO and small number of executives	None	Operating Reviews	Sell to the CEO.

Figure 16.2: Idea sales by type of enviornment

Dorian Simpson

Communicate and follow up

If a prospective customer needs more information or wants to hold another meeting, a great sales person is always on top of it. Don't wait weeks or send a bunch of unrelated information that just complicates the decision. Focus on the specific needs at hand to move the decision-making process forward.

Step 2: Pre-Sell Your Opportunity

Decisions don't just happen in meetings—actually, they rarely happen in meetings. People talk. They talk at lunch, over drinks and down the hallway. During these chats they form opinions about you, your reputation and your idea...and that's when they make decisions. Because of this fact, for any chance of success, you must pre-sell your idea. You want people to talk positively about you and your idea well in advance of any formal meeting. You want to circumvent people's natural reaction to hearing something new, which is usually looking for some reason why it's a bad idea. This is particularly true if it's not 100% in line with their current beliefs.

Consider what happens when an idea is not presold. Visualize a big meeting where someone presents an important new idea to a fresh audience. The presenter wraps up and asks, "What do you think?" In that split second before anyone speaks, everyone glances around the room, looking for cues on how to frame their reaction. Even when people have a positive reaction, they almost never speak until they're certain others in the room have a similar reaction; especially so if the most influential person in the meeting does not. Since the most immediate, natural, reaction to a new idea is typically negative, the first voiced reaction also tends to be negative. Once this happens, it's extremely difficult to turn the situation around. To change this dynamic, you must obtain the support of anybody influential before they start forming opinions through other channels or in front of you at the big meeting.

Set up preview meetings well in advance

To get people on your side, set up short, 15-20 minute meetings with the CFO, CMO, and key sales and technical executives. If you followed the steps in Chapter 9, you've already met with one or more of these executives to understand their needs. Go back now and share your proposal. Find out if they grasp the value of your concept, feel you've addressed their needs, and have any other objections to your opportunity. Set these meetings up with enough time to listen, respond, and rework your proposal until you've got their support.

Try to avoid the executive inquisition completely

Imagine not even having to present! If you've done a great job selling your opportunity in the days leading up to the scheduled meeting, you may not need to. On one project for a new web-based service, the innovator spent only seven minutes providing a summary of his plan and sharing his MRA before getting the green light. How did this happen? He did a great job pre-selling beforehand. At the meeting, the CEO reviewed the summary, asked if there were any objections (there weren't), and approved $175,000 for the team to move forward. This worked because the innovator had already met with each decision-maker in advance and addressed any concerns ahead of time. The CMO, for example, was concerned about the risk of acquiring customers. She wanted the web service to go viral quickly and needed a plan to make that happen. The innovator went back to the drawing board and developed some marketing tactics including an early pre-launch promotion that could test if the web service could quickly generate subscribers. This satisfied the CMO, allowing her to approve funding. Not every innovator will have this level of attention from executives, or get the same results, but pre-selling to avoid a big review meeting should be your goal.

It may not be possible to talk with everyone who'll be at the review meeting. Focus on the most influential executives. And don't forget to follow up! You can't just have one meeting. Some innovators take the time to meet with executives, but then don't go back to make sure their concerns have been met. If you wait until the final presentation to find out, you won't know 'til then if you have a friend or foe.

Listen for objections to resolve

When pre-selling to executives, target sections of your proposal to each individual based on his or her role and interest. It's tempting to walk through the whole proposal and share every detail, but you're not selling details…you're meeting objections. With each executive, go through the basics of your concept, then target accordingly. Marketing leaders don't need the technical details and technical leaders probably don't care about your marketing plan. Chapter 9 listed some questions to ask executives when learning their needs. Here are some additional ones to consider asking in a pre-sell meeting:

- What are your biggest concerns with this proposal?
- What concerns do you think (funding executive) will have?
- Are there any major problems that would prevent you from supporting this?
- What is the one area that I should work on improving?

And don't forget the probing questions like, "Why is that important to you?"

Finally, never be afraid to ask, "Is this a concept you would fund?" If they say no, follow up to understand why and work toward a yes. If you can get a firm yes, then they'll have a much more difficult time changing their mind at the big meeting. When that day comes, there should be no surprises as to questions that will come up and who is asking them.

Step 3: Prepare For The Big Day

Even the savviest innovators can make or break their chance for funding the opportunity proposal simply by their presentation skills on the day of the big meeting. Any decision-makers that were on the fence will make a decision based on what they hear, see ,and experience when you present.

THE POWER OF ONE GREAT PRESENTATION

Barack Obama zoomed to national prominence with one convention speech. And of course John F. Kennedy galvanized the nation toward reaching the moon with one speech.

I still remember one of my first job interviews after college. I met with a VP at NCR who was no more than 25. I had to ask him, "You seem really young to be a vice president, what is your secret?" He responded, "Well, when I just started here, I volunteered to be on a team that was investigating a new market. I was the point person and ended up giving the presentation. I guess they liked what they heard and they put me in the management fast-track program." One presentation propelled him to success.

If you're not comfortable speaking in front of people, then you may need some help. When I started consulting and training, like so many others, I was terrified of making a presentation. To overcome this fear and learn the techniques of public speaking, I attended Toastmasters International for over a year to become a "Competent Toastmaster." I can't recommend this program more highly.

Prepare like a professional presenter

What competent toastmasters and professional presenters learn is to prepare their content until they know exactly what they want to say and how to say it to get the results they want from the audience. They may look polished and natural, but no speech starts out that way. To help prepare for your opportunity presentation, here are some tips:

Memorize the story, not your bullet points. You should know your content so completely that you're comfortable giving your presentation with no visuals at all. Although Steve Jobs was notoriously meticulous about scripting every word, other presenters prefer to be more spontaneous. I'm somewhere in the middle.

Regardless of your style, if you master your story, you'll be calm and clear—two essential components of a great presentation. Whatever you do, don't read the slides! Everyone can see them and will read faster than you will speak them. Simply highlight the most important points and move on.

Practice with a live audience. I always think I've got a presentation polished and ready to go...until I try it on a real person. If you can, practice with the most senior person you know and get feedback on how confident you appear, the clarity of your story, and how compelling your concept comes across. Nothing will give you more confidence than practicing with a real audience. If you can't find an executive, practice with a trusted colleague or friend.

Nail the timing. Some years ago (when my son was little), I competed with a humorous speech on the experiences of how to manage a crying baby. My audience loved it! But my speech went over by 15 seconds and I was disqualified. I learned my lesson. Know the length of your presentation and allow plenty of time for tangents, questions, and discussion. Practice will help so you can nail the timing. You don't want to rush the final moments, particularly leading up to your MRA and the big funding request.

Step 4: It's Go Time!

So you've managed to get everyone in a room to hear your BIG idea. The executives are assembled and it's time to take the stage. Assuming you've done your homework and you're fully prepared, this is a huge moment—your opportunity proposal should be something of pride. You may not get the funding you request, but if you've put in the work to become a savvy innovator, you'll have the respect and admiration of your audience. It's go time.

Consider whether to hand out slides

Sharing your presentation slides creates a conundrum. If you hand them out, the audience reads the slides ahead of time, often without context since the slides don't tell the whole story. They might even read and not listen to you. If you don't share your presentation slides, you may be perceived as unprepared, or worse, your audience gets distracted by other work, email, or Twitter feed while you present. I always err on passing slides out. If you've prepared an engaging story and kept your presentation short and compelling, the risks of handing out slides are minimal. Besides, many people like to take notes on the slides. And they may catch key points that you didn't specifically address in your speech.

Responding to questions

You'll rarely get through a few minutes before the questions start flying. Remember, attention spans are short and executives are accustomed to watching painful presentations. When a question arises, it's tempting to say, "Wait! I address that question in twelve slides." But you'll only frustrate the person asking the question. It's better to stop and answer it succinctly, but don't let it completely derail your story. If you don't have an answer, that's OK. Admit you don't know and that you'll get the information and get back to them. Just be sure you do follow up on important questions. Few people do this. Those who do stand out as organized and persistent. This is a great way to continue the dialog with decision-makers and show that you are passionate about the concept.

Avoid heated debates

When questions and objections are raised, you definitely don't want to engage in a heated argument. You might be right (and it always feels good to be right), but you have little to gain with a prolonged debate with a more powerful adversary. The exception to this is when you know from experience that your company encourages a healthy exchange of different opinions during meetings with no adverse consequences. However, this situation is unusual. It's usually best to calmly state your case and then address the objection after the meeting if you can't resolve the issue quickly.

Quit while you're ahead

I remember one of my first big presentations to a group of industry leaders. I was requesting joint funding to do a cooperative industry marketing initiative. It was like herding cats, but after some persistence, I finally got the green light to move forward. Excited and pleased with myself, I kept talking about options for the program. As I talked, I noticed my manager was trying to get my attention. He was mouthing some words that I could barely make out. Eventually I saw that he was saying, "SHUT UP!" Once they've reached a decision, either positive or negative, accept it and plan your next steps.

Possible inquisition outcomes

Even if you've received all indications in advance that you will receive funding, this may not be the outcome. At least maybe not exactly. Following are the most typical outcomes in order of their level of success:

"Yes! How can we accelerate this idea?!" Sadly, this is rare. However, if you've done your homework well, you do have a chance of achieving this. In this situation, decision-makers are so excited that they want to accelerate the plan you've presented. They might add: "How fast can we do this?" or "Who do you need to get this going?"

"Sounds promising! Let's move forward (with funding)." Success! This is the outcome you've been working for. You've gotten most of what you want with the charter and resources to move on to the next stage. It's time to roll up your sleeves and work your MRA.

"Sounds promising, but come back with more data (without funding)." Even when you've taken all the right steps to avoid this, it is still a likely outcome. That's ok. You've probably gained the respect of your audience and have the credibility to come back with more accurate information to get funding in the next meeting.

"No...but what about this direction?" In this case, decision-makers may not like the strategy, value proposition, technical approach or other factor, and they want to redirect the effort. They may like the growth potential in the market you identified, but want you to consider a different customer problem, a different solution, or a different market segment than what you're recommending.

"We need to think about this." Decision-makers are not convinced that your idea is a good opportunity or that the risk is acceptable. This usually means the innovator didn't do his homework. It happens when key elements of the proposal are not believable or other executive needs haven't been met. You can recover from this, but you'll need to investigate the real reason behind the response.

"Thanks, but no thanks." This outcome is almost as rare as the first one. No one likes to give a definitive No! to new opportunities. They want to keep their options open. What if it has merit down the road? And also, most leaders don't want to deflate an innovator by directly saying no.

Well...that was clear as mud

Don't be surprised if after the presentation, you enter into a kind of holding pattern. Even with a concise, compelling proposal that was successfully presold, the response might still be, "We need to think about this." A savvy innovator follows up with, "I understand you need some time to think about this. When is the best time to get back with you?"

This likely outcome is another reason to have a clear, low-risk MRA to keep moving your idea forward. You should be able to say, "If we meet again in 30 days, I'll be able to show you this." Or, "I understand there are a lot of unknowns here. It sounds like the top concerns are related to the market forecast. I'd like to research this further and come back with better data." Whatever the follow-up request is, it should be targeted at the needs of the decision-makers to remove any objections.

Undesirable outcomes

Regardless of how much you prepare, things can always go wrong. Here are some of the most common undesirable outcomes I've seen:

Someone tries to hijack the meeting. Sometimes influential executives try to take over the meeting for some reason such as they love the idea, hate the idea or just want to show their power. This is a difficult situation. The best you can do is calmly direct the discussion back to your proposal. If you can't, then at least try to manage the meeting without damaging your reputation and try to regroup afterwards.

One executive is adamant it's a "bad idea." This could be a big or small problem depending on the influencing level of the executive. You should listen to their objections and work with them after the meeting to resolve their concerns.

The concept is given to someone else. Sometimes an idea has merit, but may belong somewhere else in the organization. Perhaps your executives just want someone else to take over. At this point you must check your ego and let go if that makes sense. Of course now the chance of seeing it executed to fruition goes down dramatically. This is obviously frustrating. There's no harm is asking to stay involved with the project.

Step 5: Manage The Aftermath

Regardless of the outcome, what you do next will have a long-term impact on your reputation and success. Sometimes you have to go back to people who objected, and work to understand them more

---

clearly and resolve any issues so they don't derail future meetings. These objections become the seeds of corporate antibodies you need to fight later.

Frequently you hold a review, get asked to obtain more information, hold another review, and continue in that cycle after repeated tries. It's often difficult for executives to make a formal decision until they absolutely must. If this happens, you have a choice to make. Do you move it forward under the radar? Keep driving toward a decision? Move on to the next idea? Or move the idea out of the company? If you decide to stick with your idea, assuming you've followed all the steps in *The Savvy Corporate Innovator*, your persistence and homework should eventually pay off and you'll receive your funding. Whatever you do, always think and act like a professional and follow up flawlessly.

CHAPTER 16 REVIEW—SELL THE PROPOSAL

Selling an idea requires more than a great concept. It starts with a concise, compelling, and complete opportunity proposal, that then requires a consistent selling effort. Success requires persistence, gaining early support from the right decision-makers, preparing diligently for the executive inquisition, and meticulous follow-up after the big meeting.

Savvy Success Timetable

Times will vary by project and executives' needs, but a savvy innovator should use a balanced approach to address every aspect of an opportunity. As a guideline, here are the estimated hours you should spend following these steps of your 30-Day Action Plan to sell your proposal:

Step 1: Think and act like a professional sales person	0 hours
Step 2: Pre-sell your opportunity	4 hours
Step 3: Prepare for the big day	4 hours
Step 4: It's go time!	2 hours
Step 5: Manage the aftermath	Varies widely
Total:	**10 hours**

MossBeGone
Mark's 30-Day Action Plan (continued)

Objective 7: Sell The Proposal

It's Monday morning and Mark's big presentation is scheduled in one week. He's done his homework and has a preliminary presentation that's ready to roll. It's time to make sure his presentation goes smoothly and leads to his ultimate goal—funding for his MRA.

Step 1: Think and act like a professional sales person

Mark has already been working through all the right steps to sell his opportunity. He interviewed executives to understand their needs, solicited a wide range of help from colleagues and experts, and kept his manager informed throughout the process of developing his proposal. Mark also made Tony, the CFO, an ally by including him in the forecasting effort and leveraging Betty's expertise with the financials. Now he has to put the final touches on his selling plan.

Step 2: Pre-sell your opportunity

As Mark sets out to have a few last-minute conversations with the decision-makers to pre-sell his opportunity and increase his chances to get funding, he focuses on what he knows to be Grant's (the CEO) main hot buttons:

- Does the opportunity fit with RoboCo's strategy?
- Is the market big enough and growing?
- Can the company go after the opportunity with a low investment?

Grant is going to be a tough sell. He also has a personal policy against hearing review pitches before a meeting—he preferred going with his gut reaction to new opportunities and liked seeing if presenters had done their homework.

Mark also knows that Patricia, RoboCo's CMO, will be particularly skeptical of a consumer robotic device. To learn more about her concerns and pre-sell his opportunity, Mark sends her a note:

> "Hi Patricia, I am presenting an opportunity on Monday to develop a new consumer robotics device that's considerably different than our current products. I have a preliminary presentation, but would really value your input on the marketing section before the review. I know you're busy, but do you have 20 minutes to review this with me? Best regards, Mark"

Luckily, Patricia has time to meet, and during their discussion, she expresses concerns about the market size and whether the technology will work well enough to satisfy customers. She thinks the biggest moss problems will be too difficult for a consumer-type device, and those customers who have minor moss problems won't buy the device. After Mark shares his customer interview data, Patricia softens up some on the latter point and accepts that maintenance and prevention will still be valuable. "I think you're onto something," she says, "but I'm not convinced this is big enough for us. I'll keep an open mind and see what the other folks have to say."

Mark has similar conversations with two others on the executive committee. He meets with Ted, the SVP of Manufacturing, who's concerned that the company has never done a consumer product before. And he meets with Samir, RoboCo's Chief Legal Officer, who's concerned with customer liability issues. In both cases, they agree that the first priority is to determine whether this is the right opportunity for RoboCo before working through any details.

Step 3: Prepare for the big day

Mark doesn't dare go before the executives without doing a dry-run first, so he enlists a couple of his colleagues to listen to his pitch. This helps him to practice clarifying key points and to feel more confident. He also later practices with his spouse, who helps him see when he goes too far into details and gets off track from the story.

After all of his preparation, Mark's ready to go.

Step 4: It's go time!

Monday morning in the executive boardroom. Mark confidently enters the room and counts ten executives and staff members, each doing something on their phone. Grant sits at the head of the table, with his arms crossed, looking like he's waiting to be entertained. After a brief introduction from one of Grant's staff members, Mark takes a deep breath and begins:

"How many of you have had to clean the roof on your house recently?" Mark knows that at least four of them had and it wasn't fun. Grant didn't expect this opening and was surprised when seven people raised their hand. "Was it a pleasant experience?" The audience chuckles, knowing full well that it wasn't. Now that he's grabbed their attention, Mark goes on to explain the problem and the market potential. Britney, the VP of Customer Service, asks the first question, "Do you really think there's a big enough market for a robot to clean roofs?" Mark is ready with a response, "I do. This is a big problem for homeowners. The level of frustration from the customers we talked with is very high. However, we still have some work to do understanding the types of roofs that can be cleaned and how much is spent on cleaning roofs today." Similar questions fly and Mark fields as best he can.

Trying to read the room, he's not sure exactly where he stands, but eventually gets to his MRA. "I'm asking for $250,000 to investigate this market further and work with a vendor we've identified on a rough prototype. With this money, we'll have a much better idea on the size of the market, the types of technical problems we're facing, and how well the product will be able to perform."

There is a brief silence. Everyone glances around the room looking for reactions. Grant had not spoken throughout the 20 minute presentation. Finally he says, "Ok Mark, this is very interesting. Nice job on the presentation. You clearly put a lot of thinking into this. Here is what I'd like to do. I'm not convinced this is big enough for us, but I believe you're on the right track that consumer robotics is a trend we can't ignore. I'd like you to keep investigating this opportunity and research the top 10 companies in this space." Mark can't help but grin, "Great! What do you think about developing a prototype?" Grant thinks for a moment then says, "I don't think we're ready for that. But I do want you to go deeper into how effective a roof-cleaning robot could be. I need better numbers on the types of roofs and how many homes actually have this problem. How can we do that without investing the $250,000?" Luckily, Mark has thought of this. "One of the analysts I spoke with gave me an estimate of $25,000 to research roof types further and was confident he could give us good numbers on this problem. We can also do a technical proof-of-concept with the prototyping vendor on just a portion of the device to test its ability to clean. That would be about $80,000. So I think we can meet your needs for about $115,000." "Ok," Grant says, "Update your proposal with the specific milestones and how you'll use the money. If it looks good, we'll explore this further. Anyone have objections?" None are raised and Grant closes the discussion.

Step 5: Manage the aftermath

Mark leaves the meeting with mixed emotions. He's happy for the kudos he received on his presentation, but the outcome wasn't exactly what he wanted. He quickly sends an email to each person he had met with previously to thank them for their support and get their thoughts. Universally, he hears, "That was the best response you could have hoped for." With that, Mark starts strategizing his next steps.

We'll follow up with Mark and see how his MossBeGone project plays out in the next chapter—Life after 30 Days.

(To be continued…)

PART III CONCLUSION

The goal of Part III was to offer a step-by-step plan for developing a concise, compelling, and complete proposal for a new product opportunity. Ideally, a savvy innovator should be able to follow such a plan using the specific actions and tools we addressed in approximately 80-100 hours over 30 days. The story of Mark and his Moss-BeGone new product idea demonstrates exactly how to complete the steps and accomplish this goal.

Although we focused on a new product concept that would be sold to external customers, the 30-Day Action Plan can just as easily be applied on an idea for entering a new market, taking an innovative new distribution approach, or selling a new IT system inside the company. The steps and approach would be similar along with the importance of targeting executive and customer needs. However, your approach to risk, forecasts and other aspects of a specific opportunity proposal would need to be modified for each unique situation.

As we move to Part IV, titled, Life After 30 Days, we'll take a step back and look at the broader world of corporate innovation and the mechanisms that make it work for people who have solved the innovation puzzle.

PART IV

LIFE AFTER 30 DAYS

Navigating The Corporate World Of Innovation Until Your Idea Becomes Reality

Chapter 17: Living In The World Of Innovation
       Managing the Three Unique Ecosystems

INTRODUCTION TO PART IV

The world of innovation is just like earth...there are lots of wars but also a dream of world peace.

So far we have focused on the first 30 days in the life of an idea. The initial period in the life of anything is obviously critical to its long term success, whether it's a human, rocket or pumpkin. But after this initial period it's equally important to provide the right conditions for long-term development until it has realized its potential in the world. For an innovation opportunity, long-term success is measured by its successful transition into normal business operations.

As new opportunities gain momentum, they require more people, more management attention, and more funding. These resources will likely be at the expense of current products and projects. In some, more enlightened organizations, these conflicts are understood and overcome as a natural part of managing a business. While in other organizations, a maturing opportunity may cause intense political issues as more antibodies emerge with a desire to maintain the status quo or take a different path.

Part IV addresses the challenges of managing an opportunity after successfully gaining initial funding. We'll look at the broader picture of corporate innovation and break it down into three unique systems that must work in harmony. You'll see what an innovative company looks like and how you, a savvy innovator, can successfully guide your opportunities through the world of innovation until they can survive in the day-to-day business operations.

Chapter 17

LIVING IN THE WORLD OF INNOVATION

Managing The Three Unique Ecosystems

The world of innovation is easy to navigate if you can manage the border crossings.

Rate Your Current Skills

- How familiar are you with all the elements of an innovation culture?

- What does it take to manage the gaps of weak or non-existent innovation processes?

- How do you successfully navigate tough environments and change directions if necessary?

- What mechanisms are necessary to keep an opportunity on track?

- How do you successfully transition an opportunity into normal business operations?

Introduction

Now that you've received the first round of funding to move your idea forward, your journey is well underway in the exciting world of corporate innovation. You've also earned the respect of executives and laid the groundwork for long-term success. Celebrate! But now also prepare yourself for new challenges and obstacles. As you continue developing your opportunity, you must diligently apply all the skills and tools discussed throughout this book.

For each phase, the same framework of planning, executing and selling is consistent. However, you'll need new allies, new skills, and even more determination to cross the next major hurdle of moving your opportunity into the normal business operations of your company.

The World Of Innovation Is Really Three Unique Ecosystems

Business literature is full of guidance on how to build a culture of innovation in any organization. But success is not about creating one unified culture. As you gain more experience, you'll discover that the world of innovation is made up of three distinct ecosystems.

The three unique ecosystems of the innovation world are: ideas, opportunities, and operations. Figure 17.1 shows a corporate innovation map of all three. It also shows where you are on your innovation journey after successfully developing an initial opportunity proposal and receiving your first round of funding. Let's take a look at these three ecosystems and some of the elements a savvy innovator must master to guide an opportunity through to its ultimate destination as part of normal business operations.

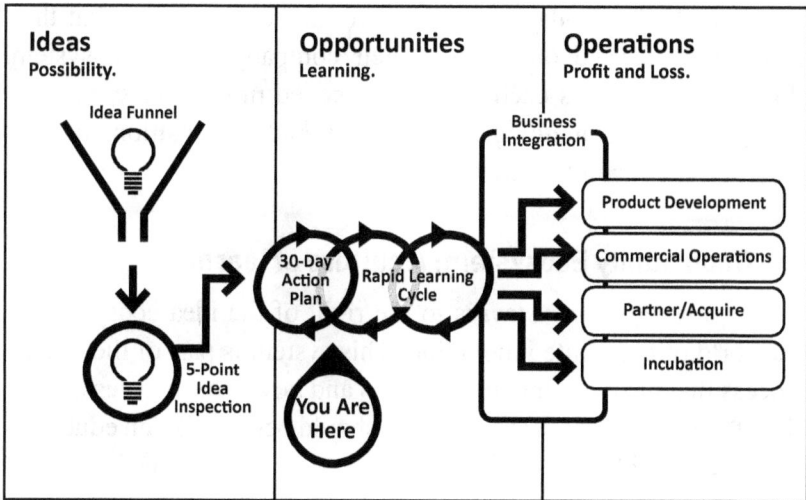

Figure 17.1: The world of innovation—three unique ecosystems

The Idea Ecosystem: A culture of possibility

The *idea ecosystem* is on the left side of the world of innovation. Since all innovation starts with an idea, this too is where innovators must start. While anyone can be part of this system, it's usually the realm of technologists, marketers, designers, sales leaders, and other creative people.

Companies with successful idea systems have created a culture of creativity, ideation, and identifying possibilities. The system encourages and searches for new concepts. Terms like 'mind maps' and 'lateral thinking' have been widely adopted, and it's OK to get a whack on the side of the head to spark new thinking. This culture makes it acceptable for participants to think differently and use words like 'risk' and 'failure' without being shunned.

Some of these companies have people fully responsible for creating an ecosystem of ideas. Although for most companies, no one has direct responsibility, but everyone is encouraged to participate. And in the worst companies, there might actually be a system that inhibits searching for ideas, and also attacking the ones that get generated.

When we read about Bob (wireless heart monitor), Mark (MossBe-Gone), and all the other innovators in this book, we saw that their initial ideas started out as part their company's idea ecosystem. However, as soon as each innovator needed time, money, or other resources to do something about it, they had to transition successfully into the second ecosystem.

The Opportunity Ecosystem: A culture of learning

The *opportunity ecosystem* is to the right of the idea ecosystem in the world of corporate innovation. This system is part of the overall process that provides the mechanisms and governance necessary for ideas to evolve, grow, and mature. It should behave like an education institution for ideas.

Companies with successful opportunity systems have created a culture of learning, fast failure, and managing complexity. Smart opportunities advance quickly while lesser opportunities are encouraged to leave the system. Those responsible for this system are skilled at managing innovation projects and helping opportunities transition to become respected contributors to the operations ecosystems. In these companies, some people are dedicated to the responsibility of building a strong opportunity ecosystem, or at least support their innovators during the opportunity phase of a project. Other, less supportive companies let opportunities starve or get so beaten up, that they can barely survive even if they do make it into the normal business operations.

The Operations Ecosystem: A culture of profit

The *operations ecosystem* is the largest and most dominant of the three and where everything is tracked, managed, and improved upon. The mantra is, "If you can't measure it, you can't improve on it!" This is not necessarily bad, but the culture is often at odds with the other ecosystems, where adhering to this mantra is difficult. However, the nature of business requires that every opportunity must find a home in this system and adopt its culture of profit.

Many people work almost exclusively in the operations ecosystem. It's the domain of finance, sales, and production people...and even most executives. Unfortunately, a lot of these people fear that innovation may disrupt or threaten the system they have worked so hard to refine.

In the best systems, people responsible for operations success (especially executives) realize they need ideas and opportunities for long-term success. They actively participate in the ideas and opportunities ecosystems and take responsibility for helping remove fear and integrating ideas into their operations. In less innovative operations environments, ideas and opportunities are feared as threats to be quickly subordinated to short-term priorities or extinguished altogether.

Boundaries Are Always Problematic

As you might expect, most major conflicts occur at the borders of the ecosystems with significantly different cultures. The world of innovation is no different. The boundaries between the three systems and the nature of the threats to each culture create the most problems.

The first chance for conflict usually occurs when transitioning ideas into opportunities. For example, in the idea ecosystem, it's OK to have a culture of chaos. Sometimes chaos is even desirable to spark new, radical thinking. However, this chaos can't be tolerated in its neighboring system of opportunities. A border that is too guarded and allows too few ideas through will stymie innovation, while an overly fluid border will allow the chaos to spill into the more organized systems for managing opportunities. Without a clear way to manage this first border, new opportunities will not live up to their potential in the world of innovation.

The second border that sparks conflict is between the flexible ecosystem of opportunities and the rigid ecosystem of operations. Because of the perceived danger of these conflicts, this border may be overly protected by those who manage the operations ecosystem. Managing this border takes exceptional skills since integrating into business operations requires a lengthy process of indoctrination and assimilation. Companies that successfully manage this border watch opportunities grow into profitable components of the operations eco-system. Unsuccessful companies waste energy on prolonged skir-mishes and create disgruntled citizens on both sides of the border.

Crossing The Border Into Business Operations

Innovators must be acutely aware of their corporate environment and be able to strategically navigate each case. Figure 17.1 shows the four possible paths to successfully integrate an opportunity into the business operations of a corporation. They include:

Product development processes. The easiest integration path for enhancement or new product ideas is to flow naturally into a company's New Product Development (NPD) process. Most companies have some form of NPD system where product ideas can be evaluated in the earliest stages as the entry point into the process. However, from my experience, without a broader in-novation strategy, most NPD processes are poor at identifying new opportunities or helping them mature. Many companies try to wedge opportunities into their NPD before the opportunity is ready. Unfortunately, this is often done to create a harsh filter in the system rather than mature a fledgling opportunity. A savvy innovator understands the dynamics and limitations of the in-ternal NPD process and chooses the right time and approach to integrate an opportunity into the process.

Commercial operations. Another path, particularly for new ap-proaches that don't require product development efforts, is to integrate the opportunity into the company's sales, marketing, manufacturing or other commercial operations. Unless your company has already developed a mature innovation process that can handle these types of ideas, most companies have no official process for managing these opportunities. For example, let's assume your idea is to create a new interactive online ex-perience to demonstrate your products. This opportunity would support sales for every product. But since it's not a new product idea, it wouldn't fit with an NPD process, so it would likely be debated as part of an operations review. Because these reviews typically focus on sales targets, production forecasts and short-term goals—not innovation—the idea is only briefly discussed then usually postponed or shrugged off. Subsequently it sits until the next available planning review, like a strategic planning event…and of course by then, is forgotten. This might also be true with ideas for new markets and any new approach to doing business. Savvy innovators must be able to move ideas forward when no formal forums exist to review and manage an opportunity, and they must be able to help the opportunity find a home in the company as well.

Partner or Acquire. A popular path that can't be ignored is transforming an opportunity into a search for a partner or acquisition. This search could be for technology, marketing channels or any other desired outcome necessary to execute an idea. As a partnership, it could be with another company, university, development lab, supplier, or any entity that can help lower the risk of a new opportunity. The open innovation movement has created more awareness for this path and made it a more acceptable option. The "not-invented-here" mentality in technology has largely eroded in favor of using internal skills and resources since many companies acknowledge that the world of technology is moving faster than they can keep up with. Taking this path successfully is dependent on a company's culture and past experience with partnerships and acquisitions. Savvy innovators are well aware of this history and manage the risk and expectations accordingly.

Incubation. The last possible path to integrate into business operations is to incubate the opportunity as a startup business within the larger corporation. This idea of *intrapreneurship* has gained some popularity, but because it's fairly uncommon, it hasn't yet acquired a very large following. Sometimes this path is the obvious choice. For example when large opportunities must be incubated because integrating it into the business would be far too politically and operationally risky to integrate by any other means. The *New York Times* digital edition, the IBM PC, and Kodak's digital camera division are all iconic examples of businesses that were incubated inside the parent company under a separate operating culture. Opportunities had freedom to grow while leveraging internal resources. I've seen success with much smaller incubation endeavors, but in any case, it takes enormous innovator and management skills to handle these intrapreneurial enterprises.

The caveat of treating a new product or business opportunity as an incubated startup is the incredibly high risk. Executives realize that any startup attempt has a high risk of failure and that large companies cannot typically act like venture capitalists. VCs are fine if seven out of ten startup companies in their portfolio fail as long as a couple of them breakeven and at least one is a home run. Incubating a new opportunity inside a corporation is an exciting path for innovators, but unfortunately, it's one that most executives, cannot (and often should not) accept. Savvy innovators understand the high risk of incubating a startup. They think carefully before recommending this path and are flexible enough to consider other lower-risk options.

These four paths are not mutually exclusive. An idea might lead to opportunity to acquire a business and then switch paths to be incubated inside the company until it can be integrated into operations. Regardless of which path an opportunity takes into business operations, a savvy innovator anticipates the challenges of that transition and carefully navigates the environment accordingly.

Savvy Innovators Thrive At The Borders

The savvy innovator focuses on moving an idea through the ecosystem borders until it is integrated firmly in the ecosystem of opportunities. This transition requires unique tools like the Five-point Idea Inspection to quantify the idea's worthiness of becoming an opportunity. You also have to pass the executive inquisition where decision-makers guard the borders and pepper you with questions. Once you make the transition into a funded opportunity, you must stay there until your opportunity is ready to integrate into normal business operations via one of the four strategies. Each phase along this path will take more resources and likely impact other projects. You should expect more antibodies to appear since careers and reputations are going to be threatened. Even when you do all of the right things, a new opportunity that threatens the status quo can easily lead to serious political confrontations.

WHEN POLITICS THREATEN REAL INNOVATION

A consumer device company that focused on education-based technology for children was facing sagging profits. They needed a new, exciting opportunity to spur growth, so they brought in Kim, a successful intrapreneur and entrepreneur, to lead the effort. After much research, analysis and internal selling, Kim focused attention on a new kid-friendly tablet with corresponding applications. Grant, the CEO, was excited after seeing inspiring prototypes and positive customer research results. He approved funds to develop the new product line. Kim started forming teams, hiring new talent, developing the software and planning for the launch. But things got ugly as other powerful executives got wind of the pending products. They attempted to influence Grant to kill the project by saying, "That's not what we need! We need this!" and "I could use the resources better than that!"

Kim became frustrated as she tried to manage politics while executing her plan and maintaining the morale of her team. She'd had enough when another business unit went so far as to copy key features of the new devices in order to directly compete. Kim was not willing to let nine months of her time and $4.2 million of investment go down the drain. After getting nowhere with the rival business unit manager, she approached Grant and said, "This won't work. We need to coordinate roadmaps with the other groups and develop a unique brand strategy or we'll just confuse customers." Grant had hoped the teams could work together, but saw things were out of hand and agreed to set things right. Grant called for a review of roadmaps and strategies, and eventually, after a series of meetings and a few heated exchanges, the situation was corrected, but not without casualties. The rival business unit manager not only backed off his planned product, but resigned from the company three months later. Fortunately, Kim's new product line was launched to great reviews and nice profits.

As a savvy innovator, Kim tried to play the corporate game, build relationships at all levels of the organization, and work the innovation process. But ultimately she had to play her last card…the relationship she had built with the CEO. Not every innovator can play corporate politics at such a high level; even fewer can win. But this situation is all too common at every level of a company.

Throughout this process, you can certainly expect to step on toes, make some people uncomfortable, and engage in some intense debates before others will accept the changes that ultimately come with innovation. A company's long-term innovation success comes down to how well it can transition a fragile business opportunity into business operations. Your success as a savvy innovator comes down to how well you manage your company's environment and processes (or lack thereof) when leading this transition.

Managing the world of innovation

There is no one perfect culture or method for successfully managing a harmonious world of innovation. Companies with mature innovation processes have figured out what works for them, but you can't say that Apple, Google, 3M, and Proctor & Gamble all have the same innovation processes or culture. Every company must find its own path to successful innovation based on its own goals and situation.

The most successful ones, like the Opportunity Engines we discussed in Chapter 3, have instituted four primary practices to successfully manage each of the three unique ecosystems—ideas, opportunities and operations—and the borders between them. The names, methods and tactics of these four practices are different, but their purpose remains constant:

Activities to capture and filter BIG ideas. Even if a company doesn't have formal ideation activities, it should have some way of capturing ideas. Successful companies generate ideas from sales meetings, customer research, exploration of technology or just surfing the web. By having a method to capture and respond to those ideas, the company opens up the natural pipeline to continue percolating more ideas. Being able to quickly filter ideas may seem simple and trivial. But, it's never easy to select an idea before an analysis has been conducted. Good innovation processes allow worthy ideas to get some attention and bad ideas to get killed quickly before they suck up resources and time. For a company to manage innovation, it must first develop a practice for idea filtering. Review the 30-Day Idea Five-Point Inspection model in Chapter 2 as one option to identify the most promising ideas.

Methods to manage opportunities as critical projects. If an idea is big enough to deliver potential results for the company, it's important enough to call it a 'project' and assign milestones, leaders, and resources toward its execution. The 30-Day Action Plan focused on forming projects for the first 30 days in the life of an idea. After first-stage funding is granted, you move into subsequent phases of opportunity development, called Rapid Learning Cycles, as shown in Figure 17.1. Each cycle ranges from 30 days to three months and follows a similar framework of planning specific steps and working with executives toward acceptable milestones. Success means the project phases don't end until an opportunity becomes integrated into normal business operations.

A mechanism to form opportunity teams. Early-stage idea investigation can be handled by one person, but the real work involved in developing new opportunities almost always requires a cross-functional team of focused, skilled team members. When a BIG idea looks like a real opportunity, successfully innovative companies will quickly form a team to execute on that idea and provide tools, resources, time, and freedom to innovate. It's never easy to take valuable time away from other projects, but innovation teams not only supply a steady stream of opportunities, they also offer great learning and leadership experiences for team members.

Forums to manage the opportunity portfolio. In a big picture sense, multiple opportunities should always be competing against one another for resources and attention. Each opportunity will have its own potential and level of risk and will also evolve at its own pace. A company should treat this variety of opportunities as a portfolio, where each one's progress and potential contribution is tracked, discussed, and course-corrected. Opportunity Engines use a dedicated forum, comprised of tough decision-makers, to fund, kill, or redirect innovation initiatives. Without this governance forum, new opportunities can easily drift or fester inside a corporation. However, an opportunity review can't be managed in the same way as a standard operations meeting. Opportunities must mature before they're ready for entry into product development processes or any other path into business operations. Success requires flexibility, a new level of coaching, and acceptance of calculated risks to create the desired innovation environment.

If any one of these practices is missing, successful innovation can be severely hampered. A lack of quality ideas leads to an Opportunity Siphon environment, indecision creates an Opportunity Clog, and ideas that run amok become Opportunity Tornados. Yet, when these four practices to manage the world of innovation come together, they combine to create an Opportunity Engine.

Where Does This Leave You—A Savvy Innovator?

You likely won't be able to control any of these four practices. Unless you're already employed by an Opportunity Engine, your environment most likely is missing one or more of these key practices. It's your job now, as a savvy innovator, to determine which practices are in place, which ones you can leverage to the best of your ability, and how you can navigate around any missing practices to move your ideas through. If you're in a position of leadership, you may be able to fill in the missing gaps and spearhead an effective innovation process starting by creating a great opportunity management environment.

SAVVY IDEA: IF THE IDEA IS THAT GOOD, QUIT YOUR JOB

After graduating from Northwestern University, I was interviewed by an Apple executive who was one of the founding members of the MacIntosh team. After a few warm-up questions he hit me with this, "So you have a great idea for a new product you think Apple should develop. What do you do?" I took a few seconds to think about it, then answered, "I'd share the idea with my manager and see how we could get it going." "That's great," he said. "She doesn't like it. Now what do you do?" I thought some more, now considering the business side. "I would research it further to see if there was a market and try again." "Ok," he responded. "She still doesn't like it. Now what?" This continued for a few rounds as I worked every angle to overcome objections. His question kept coming back. "That didn't work! Now what?" Eventually he let me off the hook and we moved on to a different line of questions.

Months later the answer hit me…*I'd take my idea outside of Apple and make it happen!* This Apple executive had wanted me to be a product leader who was so passionate about my ideas that I wouldn't let the "system" stop me from execut-

---

ing. Few corporate environments will actively support your desire to take your BIG idea outside the corporation, but if you believe strongly in its potential and can risk being an entrepreneur, it's worth exploring. My disclaimer is that you must also consider all of the legal and ethical ramifications of this approach.

CHAPTER 17 REVIEW—LIVING IN THE WORLD OF INNOVATION

There is no perfect culture or process to manage the complex world of innovation. It's a world comprised of a complex set of multiple ecosystems that can be summarized as a system of ideas, opportunities and operations. Ideas must successfully traverse each system until they can successfully integrate into normal business operations. You, a savvy innovator, must thrive in any environment and realize that each ecosystem has a unique sub-culture that must be understood and managed. You must pay close attention to the challenges of managing the borders.

Five Savvy Success Strategies

As you hone your skills, practice these five strategies while navigating the corporate world of innovation.

1. **Manage opportunities using Rapid Learning Cycles.** The concepts, methods, and steps we've explored for the first 30 days of your idea are no different for the next 30 days or any other period of a Rapid Learning Cycle. Each subsequent stage of opportunity development requires the same attention to meeting customer needs, finding relevant data, and addressing executive concerns. Of course, as the opportunity progresses, you'll be expected to produce more accurate forecasts, clearer tactics, proven technology progress, and advancement on all fronts to get more and more funding. That's just business.

2. **Stay passionate, but stay objective.** As you continue to work on any new opportunity, it's natural to get more attached to the concept. This passion will help you overcome obstacles and accept the necessary risks. But it can also blind you to the reality of when it's time to stop or redirect your efforts. Always keep an open mind and stay objective at each stage of opportunity development. In addition, any sunk costs already expended should not be factored when considering the investment of time and resources going forward.

3. **Show early progress and results.** Every new opportunity starts with an incredible amount of risk. Executives probably took a leap of faith supporting the idea. As time passes, you'll continue to gain or lose support based on the latest perception of risk. Innovators who can systematically lower risk by showing consistent progress will gain increasing support; those who don't will lose support. Remember also that while technical progress is important, it's usually more important to show progress with potential customers and reducing market risk.

4. **Communicate. Communicate. Communicate.** As every public relation expert will tell you, the best way to overcome a crisis is through consistent and authentic communication. Keeping problems hidden or waiting to communicate until you have "good news" is a recipe for rumors and innuendo about your project. Consistent one-on-one communication with key executives will prevent any surprises that can quickly derail support.

5. **Stay connected to the corporation.** Once you're well underway, you may find yourself leading a small team or working by yourself on the project full time. Given the intense focus and time required to succeed, it's easy to insulate yourself and get disconnected from the main activities of your company. However, don't forget you still need a wide range of support from the company's experts and resources. So stay involved with what's going on beyond your project because ultimately, your opportunity needs to find a home in the same corporate structure as every other product and service offered.

MossBeGone
Mark's Life After 30 Days (conclusion)

It has been fifteen months since Mark first faced the executive inquisition to request resources for his MossBeGone roof-cleaning robot. He received $115,000 to investigate the development of a prototype and refine his forecasts by working with an industry consultant. Today, however, Mark is enjoying a video of RoboCo's new snow removal robot going back and forth on a driveway at a house in Vermont. Let's take a look back at what happened over the previous year.

Mark's second formal review meeting was held 90 days after his initial proposal. To prepare for the second review, Mark followed many of the same steps outlined in the 30-Day Action Plan meeting with executives, conducting research, etc. He even invited executives out to the parking lot where he set up a section of a moss-covered roof. He proudly watched as his early prototype whirred back and forth along the roof fragment scrubbing it clean. Mark was pleased it didn't fall off even once. The executives smiled and talked as the robot did its work. They were clearly impressed.

Back in the boardroom, Mark quickly updated everyone on the technical progress his team had made. Then he showed the financial estimates. He had new forecasts and was able to get more accurate numbers on the types of roofs, the regions around the world where people routinely maintained their roofs, and other factors related to trends in retail sales of consumer robotics. Mark was not dramatically off from his original estimates (although they still lacked enough data to estimate an accurate take rate). His consultant projected that sales would likely be lower in the United States than Mark had estimated, but would be much higher internationally.

Grant, the CEO, interrupted Mark's presentation, "Mark, we've been impressed with the work you have done on MossBeGone. You've made tremendous progress and have done everything we asked." Mark heard a 'but' coming. "But I think there is too much risk for us to develop a consumer robot from scratch. However, you did convince me that consumer robotics is a growing market that we should get into. I've directed our Mergers and Acquisitions group to look for a company that already has traction in the market and the R&D skills to develop new products."

Mark was disappointed, but understood the rationale. He knew MossBeGone was a big risk. As they wrapped up, Grant shook Mark's hand, "Nice job. I've asked the SVP of Mergers and Acquisitions to put you on the due diligence team to look at the companies we identify. I look forward to your thoughts."

After several months of due diligence and negotiation, RoboCo agreed to buy a small Swedish company that was a young but aggressive player in consumer robotics. One of its products was the SnowQueen RX, an autonomous snow-removal robot that had been part of Mark's original vision. Mark became the VP of Program Management for its next generation of snow-removal robots. Although a moss-eating robot isn't on the roadmap—yet—Mark has high hopes.

MossBeGone Epilogue

Typical of many innovation efforts, the outcome for Mark's opportunity was not what he'd initially proposed. It rarely is. Ideas morph into new ideas, product concepts become different products, and ideas for technology or new approaches lead to acquisitions or partnerships. While Mark's idea for a moss-eating robot didn't get developed exactly as he planned, he did inspire RoboCo to expand its thinking and move into a growing and exciting new market area. This should be considered a big success for Mark. Not only did he get a promotion to vice president, but his brand as a savvy innovator has never been stronger.

PARTING WORDS

As marketers know, products come and go, but great brands remain. As an innovator, your ideas will come and go, but your brand as a savvy innovator will follow your career at your current company and into future ones. By continuing to have BIG ideas, being aware of your environment, understanding executive needs, and continuing to build your own skills, your personal brand as an innovator is bound to grow and open up unlimited opportunities.

I wish you the best of luck.

INDEX

Symbols

P

Q

R

ABOUT THE AUTHOR

Dorian Simpson teaches corporations, teams and individual innovators how to thrive at the front end of innovation. Starting his career with IBM, he climbed the corporate ladder at various companies from ATT, Motorola, and startups to become an executive in product management, sales, engineering and general management—always focused on bringing innovations to life. With hundreds of innovation projects over 20 years, Dorian has refined the tools and concepts found in *The Savvy Corporate Innovator.* Through consulting and training, he's inspired thousands of product professionals and helped them transform into effective innovation leaders responsible for billions of dollars in new products. His clients and workshop participants appreciate his ability to balance the conceptual world of innovation with the tactical world of business execution. Dorian has also trained product leaders in China since 2007, where his programs consistently sell out. He has a BSEE from Northwestern University and an MBA from the University of San Diego. He currently lives in Portland, Oregon.

www.ingramcontent.com/pod-product-compliance
Lightning Source LLC
Chambersburg PA
CBHW060320200326
41519CB00011BA/1783